50 Plants that Changed the World

50 Plants that Changed the World

STEPHEN A. HARRIS

First published in 2015 as *What Have Plants Ever Done for Us?* by
Bodleian Library Publishing, Broad Street, Oxford OX1 3BG

www.bodleianshop.co.uk

ISBN 978 1 85124 652 6

Publisher: Samuel Fanous
Managing Editor: Susie Foster
Editor: Janet Phillips
Picture Editor: Leanda Shrimpton
Cover design by Dot Little at the Bodleian Library
Designed and typeset by Lucy Morton in 12 on 16 Fournier
Printed and bound by Livonia Print, Latvia, on 130 gsm Munken Pure paper

British Library Catalogue in Publishing Data
A CIP record of this publication is available from the British Library

Contents

Introduction

> [T]here is no Nation so barbarous, but that it can boast of some Ingenuity, and some Invention or other to gratify an inordinate Appetite.
>
> Thomas Townsend, *The History of the Conquest of Mexico by the Spaniards* (1738)[1]

While we are aware that plants are fundamental to all life on Earth, that we are dependent on them for the very air we breathe, we are too often inclined to think of them in personal terms as accessories rather than main-stage players: the components of a garden, or garnish on a dish of more interesting fare. Yet plants pervade every aspect of our lives, no less so in a modern urban setting than in the wilds of a rural past. As I write this, I have been awake for about three hours. In that time I have consciously come into contact with products of about forty different plants, half of them featured in this book, and in the course of the day I might expect to double this number. This is not just because my work as a plant scientist means I'm 'messing about with plants' all day, but because there are few aspects of anyone's everyday life that are not directly affected by plants in their many manifestations. Plants have their own tales; some are predictable, many surprising. Take my morning biscuit, for example. It has taken 10,000 years of human selection to produce the wheat, and enormous industrial ingenuity to manufacture the chocolate, while the histories behind other ingredients, including sugar, palm oil and coconut oil, are no less amazing. As for the tale of the tea which accompanies my biscuit – that involves smuggling, 'foreign devils' and war.

Tea is inextricably linked to both Chinese and English imperialism. Illustration by Elizabeth Blackwell, from *A curious herbal*, *c*.1737.

Globally, we exploit at least 35,000 different plant species,[2] so the choice of the fifty plants I focus upon is highly personal. However, when I asked ten colleagues which fifty plants *they* would have chosen to illustrate the importance of plants in Western cultures, I was surprised, and gratified, to discover they included 60 per cent of those on my list; the remainder reflected our individual idiosyncrasies and preoccupations.

Green plants are fundamental to life on Earth. Humans cannot live without photosynthesis, the process whereby plants convert carbon dioxide and water into sugars, through the action of sunlight. Despite having striven to divorce ourselves from the planet's ecological processes, we remain dependent on plant health and diversity for our food, our fuel and our medicines. Everything we do is ultimately dependent on plants – even, as you shall read, on plants that have not lived on this planet for 300 million years.

FROM FORAGERS TO FARMERS

Ten thousand years ago the ice sheet that encased the northern hemisphere for ten millennia finally melted. Slowly, humans gave up their hunter–gatherer subsistence to settle the land, cultivate the soil and domesticate both plants and animals. This was a key – the key – innovation in the development of civilization. In living off whatever they could pick or hunt, our most ancient ancestors' lives were ruled by the habits and habitats of plants and animals they found. Even the most primitive farming changed this relationship forever. Domestication is a dynamic process – through domestication, wild plants are transformed into crops with altered morphologies, anatomies, physiologies and chemistries. As we selected plants that were the tallest, the hardiest, the most prolific – whatever characteristic was perceived to be the most useful – we established in them heritable genetic changes. Eventually, they could no longer compete with other species outside cultivation and for many domesticates there came a point of divergence when they could no longer breed with their wild counterparts. Domesticated plants had become dependent on us just as we had on them.

Although the process was similar, different plants became the focus for cultivation in different parts of the world, dictated by terrain, climate and opportunity. In the Near East, domestication of grasses such as barley and wheat was one of the cornerstones for the birth of Western civilization. As the early farmers gave up a nomadic lifestyle, possessions no longer needed to be portable, food could be stored for long periods and seeds preserved between growing seasons; wealth could accumulate. As a result, populations increased, societies became stratified and governments developed; elites emerged with specialist knowledge. Among the elites were politicians and priests, artists and architects, mathematicians and scientists, engineers and technologists, farmers and warriors. Management of this ever more complex system required the evolution of bureaucracies, standardization of measurements and means of exchange, writing to record information and legal

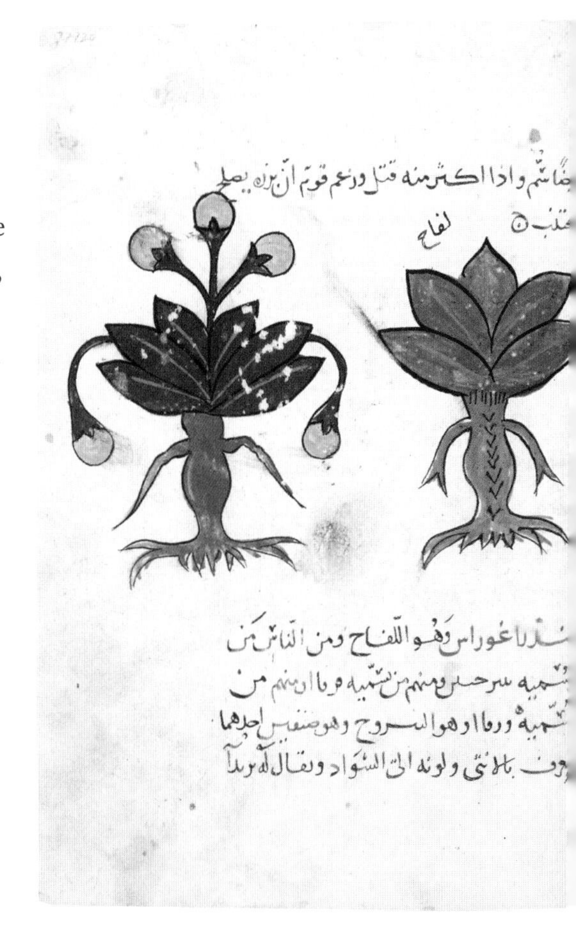

Illustrations of mandrake and nightshade from an Arabic translation of Dioscorides' famous medical manuscript, dating from 1240.

and religious systems to enforce modes of behaviour. Gradually, scientific answers were found for natural phenomena that had previously only been explained by religion, superstition and portents. Plants, and the products they produce, made significant contributions to these developments at all stages.

While it may be clear, when we pause to consider, some of the ways in which plants have influenced the way we live, the part they have played in our make-up at a deeper genetic level is less obvious. When we chose a plant that possessed particular characteristics we found useful, sometimes we also had to adapt ourselves to accept some of its other features. Wheat, that most basic and ubiquitous of grains, is just such a plant (see page 49).

Varying social, political, environmental and cultural circumstances mean people adopt different plants for their particular needs. Vegetables and fruit, from apples and beetroot to potatoes and tomatoes, are important additions to our diets and made more palatable by cooking, perhaps in palm, soya and olive oils, or washing them down with drinks made from grapes, coffee or tea. We colour our world using plant dyes and clothe ourselves in plant fibres. Paper from papyrus,

mulberry and pine records our deeds and ideas, traditionally using inks made from plants such as oak. We enliven our environments by filling our gardens with flowers such as roses and tulips. Yet these are only a few of the myriad ways in which we make use of the plants around us. They have enabled us to build, travel, explore, kill and cure.

Plants, and the biologically active compounds they contain, have always been important as medicines. Among the earliest descriptions of therapeutic plants are the 3,500-year-old Egyptian *Ebers Papyrus*, with some 150 medicinal species named, and the Assyrian herbal of King Ashurbanipal, written over 2,500 years ago.[3] The Ancient Greek and Roman knowledge of medicinal plants was distilled by Dioscorides into his *De Materia Medica* in the first century CE, and this most famous medical manuscript in Western culture formed the basis of most books and manuscripts on medicinal plants until the Renaissance.[4] We still treat many diseases today with plant-based medicines, using undiscovered plant-based therapies to justify plant conservation programmes.

Over time certain plants move in and out of favour, while new uses may be found for familiar ones. The market in yew wood for longbows is not quite as brisk as it was at the time of the Battle of

Agincourt, but the yew's bark and now its needles have a modern and vital use (see page 84).

ENTER THE ARCHAEOBOTANIST

The social, cultural and political changes of sedentary civilizations are preserved in the archaeological record. Much of what we know about our forebears' lives comes from their graves, rubbish heaps and middens. Skeletons tell us how and where people lived and died, and the quality of what they ate; the leavings of everyday life tell us what people lived on and what lived on them. As civilizations collapsed and cultures disappeared, even the holes in which they interred their dead or piled their waste ceased to exist. Sometimes, all that remains of these people's lives are a few physical artefacts and fragments of their knowledge, often about the natural world, assimilated by other cultures. Who nowadays recalls the extinct Taino people? And yet a faint echo of their life and culture whispers to us through the very words 'tobacco' and 'maize' (see pages 149 & 189).

The discovery of crop origins is more than an intellectual exercise; it has direct practical and cultural benefits. As our populations increase, we need to grow more food. At its simplest, increase in food production can occur by converting natural habitats to fields or by selection of useful crop traits. As we begin to understand how crops evolved, we can increase the precision with which we are able to breed and improve them. These skills will be essential if 9.7 billion people are to be fed by 2050.

Darwinian evolution, the unifying principle of modern biology, provides a framework to explain the diversity of life on Earth. The vista Charles Darwin and the monk Gregor Mendel (see PEA, page 65) opened to understanding evolution is based on selection of variation in natural populations, the inheritance of genetic information from one generation to the next, and adaption, over time, to particular environments.

By the mid-nineteenth century, narrow views of what was respectable for botanists to study were being challenged. The French-Swiss botanist Alphonse de Candolle recognized that academics needed to collaborate and synthesize data from disciplines as diverse as morphology, biogeography, history, philology and archaeology if cultivated plants were to be understood.[5] The few plant fragments recovered from archaeological sites, the difficulties of reliably dating archaeological remains and objectively determining plant relationships hampered him. In the early twentieth century the Soviet geneticist Nikolai Vavilov instead took a genetic approach. He reasoned that local crop variability would be greatest in areas where crops were domesticated and were in contact with their wild relatives.[6] Furthermore, local crop cultivars would be highly adapted to the places in which they grew and the conditions under which they were managed, and contain a reservoir of genes, some of which would be similar to those of the wild relatives.

The challenges of discovering the origins of the crops with which our cultures evolved should not be underestimated; the soft, leafy parts of plants leave few archaeological remains. However, over the last century, developments in archaeology, physics, philosophy of science, computing and genetics have reduced many impediments to modern investigators of crop origins.[7] Refinements in archaeological field techniques mean leaf fragments, small seeds and pollen grains are readily recovered from sites. Using modern dating technologies, minuscule samples of individual plant fragments can be dated, whilst rigorous species relationships are reconstructed using cheap, powerful computing. Vast amounts of genetic data can be recovered by DNA sequence analysis.

EMPIRE-BUILDING PLANTS

The natural distributions of plants mean few species are found everywhere. So, as people explored the world, they discovered and adopted new plants with new uses. Western civilization absorbed staples domesticated across the Old World: rice from China, sugar cane and bananas from South and South East Asia, coconuts from Polynesia and oil palm from Africa. Exploration of the New World expanded Western diets further: maize from Mexico and potatoes from the Andes, spicy chillies and exotic pineapples and tomatoes. Enthusiasm for these new crops expanded beyond the dinner table: we exploited drugs from cocoa and tobacco, coveted silk to wear and rare plants for our gardens, while the rubber tree provided one of Western civilization's most unexpectedly civilizing and destructive influences. Curiously, despite the contribution of Australian plants to global construction, in the form of eucalyptus, no mainstream Western food plants have their origins in Australian cultures.

Just as exploration had led us to these tantalizing new plants, so demand for them pushed the boundaries of exploration. New territories were claimed and protected, and sending these hard-won valuables from the ends of the Earth meant developing a secure and efficient transportation system. Timbers and grasses, such as oak, pine, yew and reed, became important for the construction of ships, weaponry and housing, while meat-based economies and animal transport systems needed fodder, such as meadow grass. Merchants became rich on trading a multiplicity of goods and services that directly or indirectly stemmed from plants. Portugal and the Netherlands were made transiently wealthy based on their domination of the sixteenth-century Eastern spice trade. As we moved and waged war, we naturally took plants with us, either deliberately or accidentally. Many of these plants were benign; a few, such as ragwort (page 235), became weeds and have played a different sort of role in the development of Western civilization.

WHERE THERE IS DISCORD...

Over 2,000 years ago, when the Greek philosopher Theophrastus wrote about plants he justified their study for economic and medicinal reasons. Researchers still justify their interest in plants in these

As Europeans explored the New World they brought home exotic plants such as pineapples, which became a favourite with the aristocracy. Illustration from Johann Weinmann's *Phytanthoza Iconographia*, c.1737.

Chillies from the Americas have transformed western cuisine. Illustration from Johann Weinmann's *Phytanthoza Iconographia*, c.1737.

utilitarian terms. However, since plants are not green animals, there are other fundamental reasons that make plants worth studying. With the current environmental, political and social threats posed by an expanding human population (in both number and weight), limited water resources, uncertain food security and unpredictable climate change, the essential importance of plants is once again beginning to be appreciated. However, our history is one of exploitation: our museums, gardens and fields are full of the evidence of us being both exploiter and exploited. Plants, and the desire for the products they produce, are intimately associated with colonization and the exploitation of people.

Evidence for our sustainable use of natural resources over hundreds of years is sparse, and may become confounded with an imagined ecological 'golden age'. George Marsh, a nineteenth-century American diplomat, was going against the received opinion of the day when he stated: 'Man is everywhere a disturbing agent. Wherever he plants his foot, the harmonies of nature are turned to discord.'[8] When stripped of their strong theological undercurrent, his ecological views are surprisingly modern.

It is inappropriate to imagine Marsh's harmonious nature as an unaltering Eden. Every part of the Earth is continually changing, adapting, evolving. The romantic image of Robin Hood harrying the Sheriff of Nottingham through a wild wood of ancient oak forests is a naïve cliché. Even had Robin existed, he would have been hunting through a highly managed landscape, transformed since the Neolithic age, perhaps resembling the New Forest as it survives today in Hampshire.[9] But the evidence of humans as a 'disturbing agent' is everywhere: in lists of species gone extinct, for example, or in devastated forests from Borneo to Brazil (see page 266).

In compiling these fifty plant profiles, a chronological approach is used, where plants are arranged by when they first became influential in Western civilization. Some, such as barley and wheat, were there at the birth of society and continue to be staples; others, such as pepper

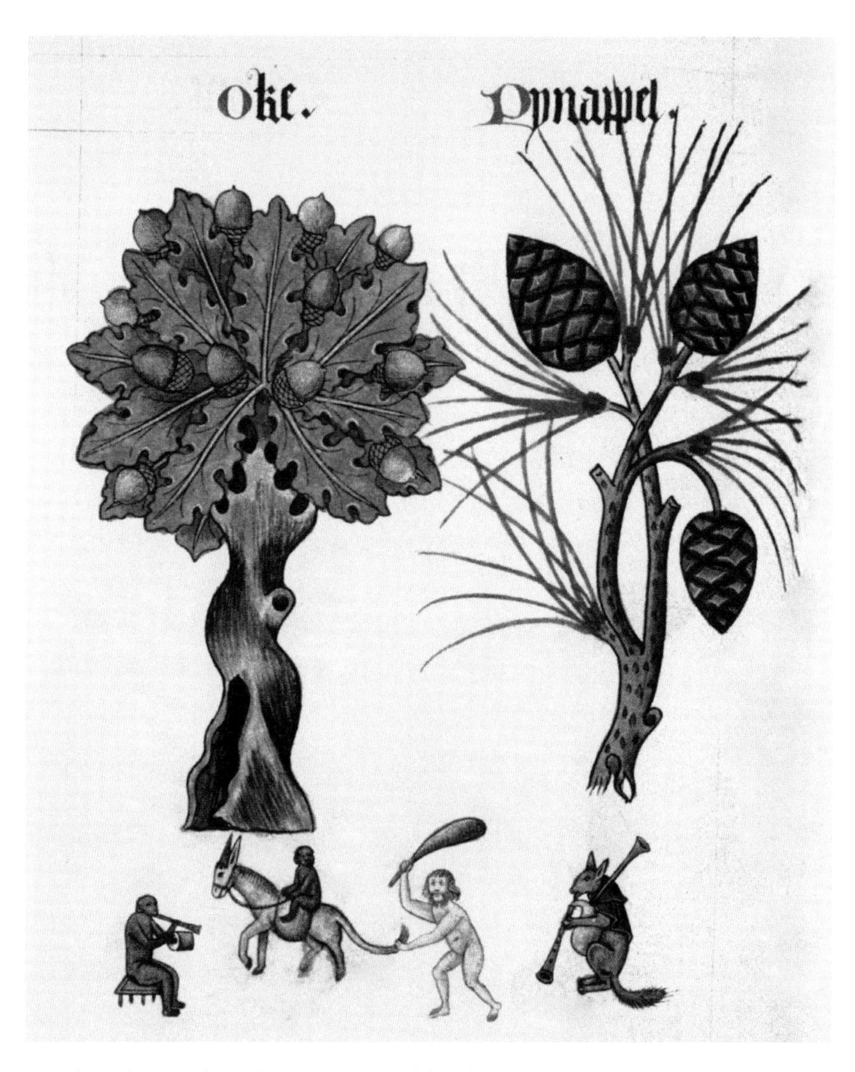

Supplies of oak and pine became essential for ship-building, weaponry and house construction in the Western world. Illustration from a Tudor pattern book, *c.*1520–30.

and nutmeg, had their reign as fantastically expensive exotics over which fortunes were made and lost but are now everyday items in many households. There are plants here that have changed landscapes, fomented wars and fuelled slavery. Others have been the trigger for technological advances, expanded medical knowledge or simply made our lives more pleasant. Plants, and plant products, are everywhere we look. Their effects on our lives continue to be profound, and often unpredictable.

A NOTE ON PLANT NAMES

Each of the chapter headings includes common, scientific and family names. There is a tendency to baulk at scientific names in favour of common names; they are familiar and seem easier to understand, but they can also be very localized and confusing, whereas the scientific name is a universal identifier. It was a Swede, Carolus Linnaeus, who created the standardized binomial naming system that we use for all organisms. Scientific names comprise two parts (the binomial): the genus name and the species epithet, followed by the name of the author who first described the plant. 'L.', which follows many of the species names in this book, identifies Linnaeus. For more on the profound difference his system made, and for a taste of what a medley of common names often attach to a single plant, see SMOOTH MEADOW GRASS (page 199).

Scientific names can also be rich sources of curious information (and disinformation) in their own right. They may relate to place (*sinensis*, 'of China'; *brasiliensis*, 'of Brazil'), habitat (*pratensis*, 'of fields') or use (*officinale*, 'used by apothecaries', i.e. medicinal; *tinctoria*, 'for dyeing'; *somniferum*, 'sleep-inducing'). Many indicate a certain characteristic (*Helianthus*, 'flower of the sun'; *fragrans*, 'fragrant'; *tuberosum*, 'having tubers or swellings'; *nigra*, 'black'; *oleracea*, 'of the garden or vegetables'). Others are used to honour the friends, patrons and colleagues of botanists (*Cinchona*, *Nicotiana*, *thaliana*). Yet other names convey a plant's qualities in a more romantic way: banana, for example, is *Musa* × *paradisiaca*, and cocoa is *Theobroma*, literally 'food of the gods'. Consequently, scientific names themselves are part of cultural history, telling us much about individual botanists and the times in which they lived.

The spelling of quoted material has in places been lightly modernized for clarity.

The plants

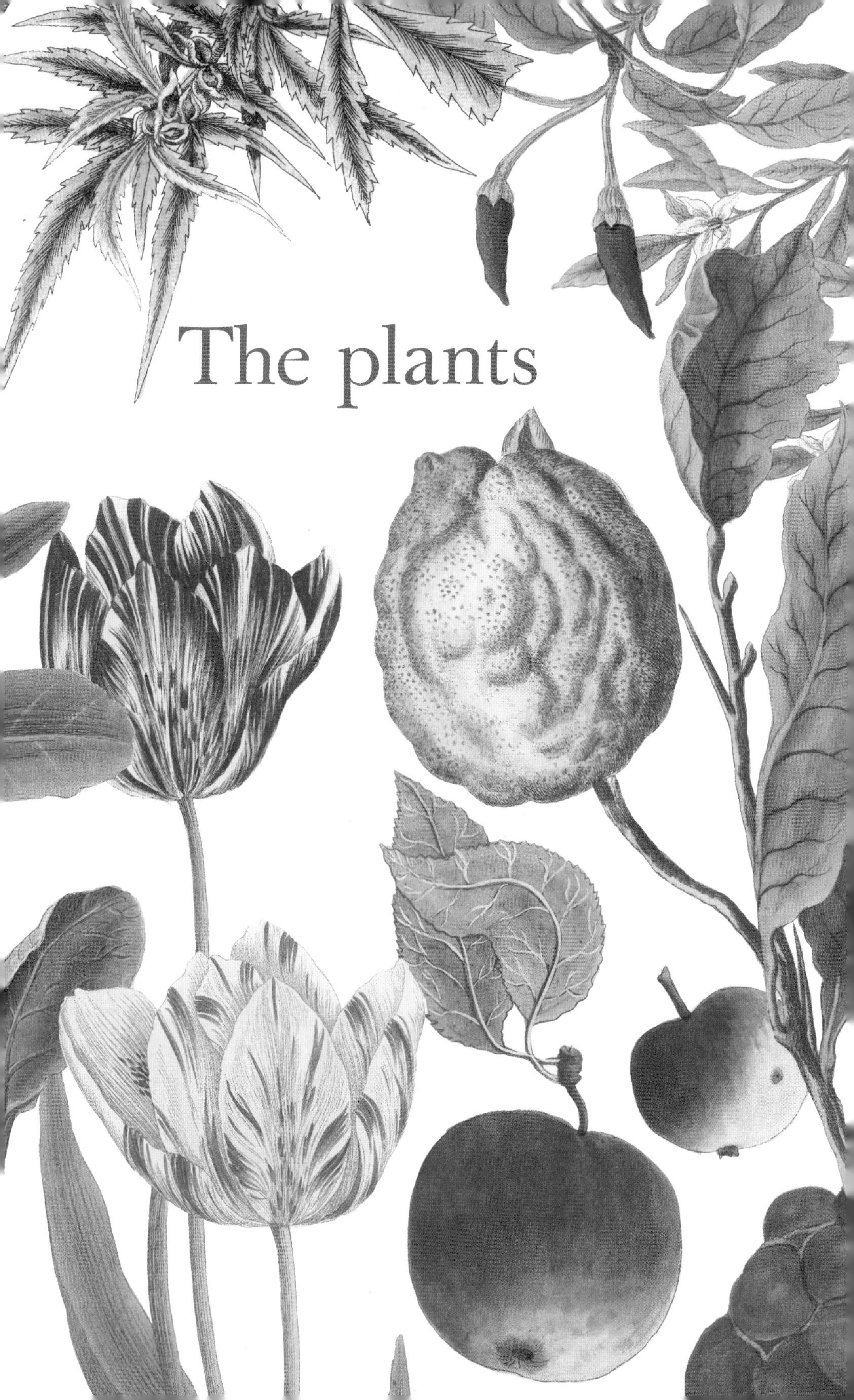

a. Higuero forte seu Ahovay Theveti.
b. Hordeum, Orge, Gersten.
c. Hordeum polystichum hybernum, Escourgeon, Wintergersten.
H.

Barley

Hordeum vulgare L.; Poaceae

BREAD, BEER AND BREWING

For millennia people have grown and traded barley, and its domestication is thought to be one of the foundations of Western civilizations. Early cultivators no doubt considered the annual cycle of tilling soil, sowing apparently dead grains and watching the crop mature a mysterious process. Little wonder that there evolved religious rituals to cajole and placate the gods when grains were sown, and to thank them following a successful harvest. Such rituals would help bind a community in a common belief and give reassurance of survival in the future – unless you were the chosen sacrifice.

Forms of worship developed and changed along with the rise and fall of societies and empires, but it is still possible to discern earlier beliefs embedded within later religious practices. In the Roman calendar 25 April was Robigalia, a festival for the protection of grain fields; the Rogation Days of the Christian calendar, when parish boundaries are beaten and prayers to bless the earth are said, today serve a similar purpose, while Christian harvest festival celebrations supplanted pagan harvest rites associated with the autumnal equinox.[10] The traditional English ballad and folk tale of John Barleycorn amalgamates Christian and pre-Christian belief, anthropomorphizing the agricultural cycle as birth, torture, death and eventual resurrection. Barleycorn figures in the sinister cult film *The Wicker Man* (1973), in which a remote Scottish community fashions bere (the locally grown barley) into the image of John Barleycorn, requiring a human sacrifice in order to become 'the life of the fields'.

The words 'bere' and 'barley' derive from the Old English cereal name *baerlic*, which is so deeply embedded in our agricultural history that it also gave us the word 'barn', or 'barley place'. But barley's origins are far from the familiar barns of cosy European or North American farmyards.

Domesticated barley is thought to have originated in the Fertile Crescent, where its wild ancestor is common. The Fertile Crescent extends in a long arc from the Red Sea north along the eastern Mediterranean into northern Syria before sweeping east through south-eastern Turkey and south-east along the Iraq–Iran border.

The first evidence of domesticated barley comes from archaeological sites near Jericho that are up to 8,000 years old. However, Tibet, where barley has been a staple since the fifth century CE, is also a major area of barley domestication.[11] The domestication of barley has resulted in two particularly important differences from its wild ancestors. Wild barleys drop their seed when ripe, whereas domestic breeds retain their ears of grain intact on the stalk, an obvious advantage for harvesting. The second is in the formation of the ears themselves. In barley flower heads, individual flowers are arranged in threes. Stacks of floral triplets are arranged on either side of the flower stalk. Within each triplet, the central flower or all the flowers may be fertile. If all flowers are fertile, then the barley produces ears with six rows of grains. If only the central flowers are fertile, each seed head forms only two rows of grains; wild barleys are two-rowed. Selection by humans has produced a wide range of different types of barley. Historically, in the absence of a clear understanding of fundamental barley genetics, a plethora of scientific names emerged, although most of these have now been merged under a single name, *Hordeum vulgare*.

In Western civilizations, barley has mainly been used for making bread, beer and spirits, and as animal feed. As a food, barley is often considered inferior to wheat (see page 49), but it endures as an indispensable crop: in 2022 more than 153 million tonnes of barley were produced worldwide, making it the fourth most important cereal. The vast majority is used for alcohol and animal feed production. Six-rowed barley is more suitable for animal feed than two-rowed, since it tends to contain more protein and less sugar. One key attribute is that barley is productive in dry, nutrient-poor environments where salinity may be an issue; wheat is much less tolerant of such conditions.

The ability of humans to brew alcohol has been argued to be a key innovation for the development of civilizations – not for our ability to get drunk, but through our ability to manipulate the natural world, and so preserve food. Fermentation converts carbohydrates to alcohol and carbon dioxide, and ultimately organic acids. The production of ethanol, the simple two-carbon alcohol that forms the basis of the type of alcohol that humans consume, needs only water, a sugar source (such as malted barley grains), an anaerobic environment and yeast fungus. Humans have manufactured ethanol for millennia; in the Zagros mountains of Iran, traces of barley beer have been found on pottery fragments at least 5,000 years old.[12] Numerous Sumerian cuneiform tablets record details of brewing techniques and barley transactions, and funerary objects from Egyptian tombs portray the transformation of barley into beer. Mesopotamian references to beer are also found in the *Epic of Gilgamesh*, an ancient poem honouring the goddess of brewing.[13]

Brewers soon discovered that beer could be preserved and its flavour enhanced by adding plants such as hops. Indeed, the traditional distinction between ale and beer is that beer was ale with the addition of hops. Furthermore, brewers exploited differences between barley types, with two-rowed barley being traditionally used for English and German ales, and six-rowed being used for American lagers. Despite the control early brewers had over the fermentation process, it was only in 1856 that the French chemist Louis Pasteur made the connection between yeast and fermentation.[14] Fermented barley grains become beer and ale, and these, when distilled, are transformed into malt whisky.

Cultures replete with their own languages, rituals and mysteries, which to the outsider can seem to verge on the fetishistic, have developed around individual drinks. But the influences on society of alcoholic drink, whether fermented or distilled, extend far beyond adherence to a particular brewer's produce, reverence for a single malt or monetary fashion. In the absence of clean water, beer is often cited as the preferred drink; even children would be fed small, or weak, beer as a safer option than typhoid- or cholera-ridden water. And then there are the political influences. In 1751 the cartoonist and printmaker William Hogarth published two prints, *Beer Street* and *Gin Lane*. Beer Street denizens are happy, responsible, hard-working citizens, while those of Gin Lane are unhappy, feckless and lazy. The messages are clear: it matters what and when you drink – English beer, good; foreign spirits, bad. But alcohol, fermented or distilled, is a double-edged sword: exchequers that tax alcohol accrue vast sums from its sale but must cope with its effects on behaviour and health.[15] The temperance movement was born in response to alcohol problems in nineteenth-century England (see TEA, page 229).

Money does not grow on trees, but the means to weigh precious metals accurately, and all that follows for economic development in civilized societies, can be found growing on cereals. In Ancient Greece a drachma contained 18 *kerata* (or carets); each *keras* was equivalent to four wheat grains. The Irish Celts measured eight wheat grains to one *pingiun*, and for over six hundred years, from Alfred the Great to the start of the Tudor dynasty, the English fixed the weight of a penny as thirty-two grains of wheat. If only barley was available, all these systems had an 'exchange rate' of approximately three barley grains to four wheat grains. When you next buy a pair of shoes, remember that your shoe size is determined by the old measure of a barleycorn.

Using cereal grains as weight and currency standards implies considerable uniformity across individual grains, and arguments abound as to whether cereal grains were ever really used as practical weights and measures. Whatever the case, cereals were evidently important enough to be associated with rigorous measurement, as the framers of English laws, aware of variations in grain sizes, tried to close such loopholes. For example, under Edward I it was specified that 'the penny sterling should weigh 32 grains of wheat, round and dry, and taken from the midst of the ear.'[16]

Barley, this apparently unremarkable grain, has propped up Western civilizations for thousands of years. In addition to its vital contribution as bread, beer and animal feed, it helped people understand chemistry and domesticate yeasts, and it played its part in transforming low-value raw materials into high-value products.

b
b
b
a
c
c
a. Mandragora foemina, Schlaffäpfel.
b. Mandragora Mas, Mandragore, Alraun
c. Mandragora flore subcoeruleo, Hundsäpfel.
5.

Mandrake

Mandragora officinarum L.; Solanaceae

THE SHRIEKING ROOT

Humans are a late arrival to the game of chemical warfare. For millions of years, plants have evolved in environments where they are under persistent attack from herbivores, pests and diseases. They have responded in many ways to such assaults, including the evolution of an enormous array of complex chemicals. These plant chemicals may be of direct interest to us since, until the late nineteenth century, they were the most effective sources of drugs in our medicine chests. Indeed, the need for modern plant conservation has been justified in terms of the potential that plants have as sources of new human medicines.

In the first century CE the Greco-Roman physician Dioscorides catalogued the identity and properties of plants he used in his practice. Four hundred years earlier Theophrastus had justified his investigations of Mediterranean plants on the economic and social imperatives of agriculture and medicine. For 1,500 years European botanical writing was guided by these two ancient authorities, although much of Theophrastus' scientific insight was overlooked. During this period the study of plants became synonymous with the recognition of medicinal and food plants. Without direct knowledge of drug sources, physicians were in danger of making incorrect prescriptions and apothecaries of being duped by adulterated raw materials.[17] In 70 CE Pliny complained that 'little by little experience, the most efficient teacher of all things, and in particular of medicine, degenerated into words and mere talk. For it was more pleasant to sit in a lecture-room engaged in listening, than to go out into the wilds and search for the various plants at their proper season of the year.'[18] Within such environments, evidence may become diluted, allowing superstition, pseudo-science and fraudulent prognostication to flourish.

For at least 4,000 years the most (in)famous drug plant in Western culture has been the mandrake. It rejoices in numerous common names: Satan's apple, fool's apple, Satan's testicles, and dragon doll, to name but a few. The plant, which is commonly found in dry areas of the Mediterranean and the Levant, is a perennial with a long, parsnip-shaped taproot. Above ground, the distinctive dark green leaves form a flattened rosette. Purplish bell-shaped flowers arise from the centre of the rosette in spring and ripen in autumn to enigmatically scented yellow berries the size of ping-pong balls. Mandrake is considered extremely hazardous because of the concentrations of tropane alkaloids (mainly scopolamine, a hallucinogen, and hyoscyamine), which interfere with the normal function of our synapses.[19] However, as the Swiss-German alchemist Paracelsus pointed out, toxicity depends on dose; the difference between curing and killing is one of degree. In the case of mandrake, low doses produce drowniness and anaesthesia and moderate doses hallucinations, while high doses kill. As with all medicinal plants, judging mandrake dose is tricky because alkaloid concentrations vary with the part of the plant from which they are extracted, the environment in which the plant grows and its stage of development.[20] The trick for the mandrake user is to know enough chemistry and biology to be able to judge an appropriate dose within a nicety.

Ancient Sumerians and Egyptians made use of mandrake and decorated their tombs with it.[21] The Carthaginian warrior Maharbal even used mandrake in an early example of chemical warfare. Charged with putting down an African uprising, Maharbal ordered wine, infused with mandrake, to be left behind as he retreated from his camp. Returning to the camp later he found the drugged rebels asleep, and had them promptly slaughtered.[22] Roman physicians used it to bring about sleep and anaesthesia, cure eye disease and induce abortions

– when medical interventions were brutal, plants that dulled sensation or consciousness were highly prized. 'Not poppy, nor mandragora, / Nor all the drowsy syrups of the world', / Shall ever medicine thee to that sweet sleep', mutters Iago, as he plots to ruin Othello's peace of mind.[23] Some researchers have suggested that chemicals in the scent of the mature fruit have aphrodisiac properties.[24] The Ancient Greeks certainly included mandrake in aphrodisiacs, and this appears to have been its function in the story of Rachel, Leah and Jacob recorded in the book of Genesis.[25] Tropane alkaloids are fat-soluble, and people soon discovered that mandrake's toxicity could be reduced by careful preparation using fats and oils, producing ointments and lotions which, when applied to the skin or mucous membranes (usually of the rectum or vagina), could realize the full hallucinogenic effects of the tropanes.[26]

These hallucinogenic properties entered into Egyptian, Greek and Roman religious practice, but it was during the European medieval period that mandrake became closely associated with the practice of witchcraft. Two close relatives of mandrake, henbane and deadly nightshade, were commonly also incorporated into witches' ointments to produce different effects, since they each contain different proportions of the main tropane alkaloids. A mandrake concoction was often given to those condemned to burn at the stake, including, ironically, women accused of witchcraft for using just the same ingredients.[27] By tradition, mandrake wine was offered to Christ as an anaesthetic at his crucifixion.

The source of much mandrake folklore is the shape of the root, which can resemble a man, and its strong associations with the Doctrine of Signatures. Classically, mandrake was reputed to shriek when torn from the ground and kill whoever uprooted it. Shakespeare's Juliet, just before feigning death, voices her fears of waking in the family tomb, with its 'loathsome smells / And shrieks like mandrakes' torn out of the earth, / That living mortals, hearing them, run mad'.[28] Incantations with swords and circles could protect the mandrake collector, but the most common means of harvesting was to have a dog: 'he who would take up a plant thereof must tie a dog therunto to pull it up, which will give a great shreeke at the digging up; otherwise if a man should do it, he should surely die in a short space after.'[29] Indeed, early mandrake illustrations usually show the root with a dead dog attached. During the medieval period, mandrake was supposed to grow where the urine and semen of a hanged man hit the ground.

All the perceived difficulties and dangers of harvesting mandrake did not prevent it being an adornment to British gardens as early as the tenth century. Rarity and the associated stories made mandrakes desirable objects; indeed, how many of the tales were concerned with safeguarding supplies for the favoured few? Mandrake roots with a strong human form were considered so powerful that judicious whittling could increase a root's market value dramatically. Furthermore, fake mandrake roots, usually carved from white bryony roots, were common across Europe as amulets and supernatural protection.

By the end of the seventeenth century, people with a practical knowledge of plants were ridiculing the old beliefs. The English herbalist John Gerard stated:

> there hath been many ridiculous tales brought up of this plant, whether of old wives, or some runnagate Surgeons of Physicke-mongers ... all which dreames and old wives tales you shall from henceforth cast out of your books and memory; knowing this, that they are all and everie part of them false and most untrue.[30]

It has also been suggested that such tales were directly responsible for the disappearance of mandrake as a medicine, as physicians and apothecaries tried to gain respectability during the eighteenth and nineteenth centuries; another factor was that better anaesthetics were becoming available.[31]

Mandrake is no longer of medicinal value in mainstream medicine, although plants remain a major source, directly or indirectly, of drugs.

Bringing a new medicine to market can take a decade or more and cost hundreds of millions of pounds, and many of today's medicines are based on chemicals that have been purified or synthesized from plants and then rigorously tested for safety and efficacy.

Mandrake mythology remains strong, with mandrake references frequently found in popular culture – for example, J.K. Rowling's Harry Potter stories. And we have found other drugs that ensnare the gullible and the desperate, apparently conferring status, sexual and financial prowess, or providing an anaesthetic from the realities of life. Mandrakes in all but name.

Beets

Beta vulgaris L.; Amaranthaceae

WILD GREENS, SWEET PROFIT

Look around waste places, car parks, beaches and cliffs on the coasts of Western Europe and the Mediterranean. You are likely to find a plant with thick, diamond-shaped, glossy green leaves and ridged stems flecked with beetroot-red pigment. This is sea beet. Despite its appearance, sea beet has been harvested from the wild for more than eight millennia; its leaves are one of the most frequently collected wild food plants around the Mediterranean.[32] In Northern Europe, in addition to the leaves, sea beet's roots were also important foods. However, more significant for Western cultures than its use as 'wild food' were the decisions different peoples took to select sea beet features and bring them into cultivation.

Sea beet was probably first domesticated in the Near East some 8,000 years ago. Initially, people selected the wild plants for their leaves, producing crops such as spinach beet and chard. The Greeks of about 1000 BCE were familiar with multiple forms of chard, many of which were gradually taken up by the Romans. Beetroots, with their swollen underground parts, started to appear about 500 BCE, although their precise origin is unknown. In Europe, during the Middle Ages, a variant with large, swollen roots was developed, and became the magnificently named fodder mangelwurzel.

Images of beet appear in wall paintings inside Egyptian tombs near Thebes, dating from the twelfth dynasty (2000–1788 BCE). In ancient times, beets were strongly associated with their presumed medicinal properties, including treatment of dandruff, chilblains and leprosy, while the sweetness of the underground parts apparently led Caesar

Beta rubra vulgaris, Poirée rouge, Rother Mangold.

Augustus to invent the verb *betizare* (effeminate behaviour). However, by the early modern period what attracted most attention was their food value. Herbals of the time are replete with descriptions of types of spinach beet, chard, beetroot and mangelwurzel, while the bright red beetroot and the coloured leaves of chard were particularly attractive and frequently used as ornamentals. The seventeenth-century French botanist Pierre Magnol went as far as to repeat the view that 'nothing is more useful in the kitchen than beet'.[33]

As might be expected from the range of forms and the length of time over which beet was domesticated, its taxonomy is complex. At its simplest, all cultivated beets belong to a single subspecies, and within this subspecies different types are recognized as botanical varieties. Further complications are added by the numerous horticultural varieties that have been bred. At the end of the nineteenth century, for example, one horticultural company boasted 5 sorts of spinach beet and chard, 21 types of beetroot and 16 types of mangelwurzel or mangold.[34] The same company also offered 12 sorts of sugar beet.

It was sugar beet that promoted beet from a cottage garden vegetable to an industrial crop. The sixteenth-century agriculturalist Olivier de Serres commented that French cooks made sweet syrups from beetroot, and about 150 years later the German physicist Andreas Marggraf showed beetroot sugar was the same as cane sugar – that is, sucrose.[35] By 1784 Franz Achard was using these observations to start selectively breeding sugar beet from 'White Silesian' mangolds,[36] and in 1801, just as Europe was pulling itself apart in yet another war, the Prussian state backed Achard to open a commercial factory for beet sugar extraction. The Napoleonic Wars saw France blockaded and food supplies diminish; sugar became scarce as access to her Caribbean colonies was curtailed. Napoleon set the French chemist Antoine-Augustin Parmentier, who is better known for his exploits with potatoes (see page 173), to investigate whether sugar could be commercially extracted from other sources. Parmentier began with obvious candidates such as grapes, but became convinced that beets were the best way to secure a sugar supply within France's borders and so break her reliance on imported sugar. By 1811 Napoleon had ordered French farmers to plant sugar beet, and by 1812 two French industrialists, Jean-Baptiste Quéruel and Benjamin Delessert, had invented a commercial process for extracting sugar from beets; in 1813 Napoleon banned the import of Caribbean sugar into France. By 1837 France had more than five hundred sugar-processing factories and was the world's largest sugar beet producer, a position she held until 2010. During the nineteenth century, sugar beet cultivation gradually spread across the temperate world, but it only became part of the British agricultural landscape after the First World War. (See also SUGAR CANE, page 215.)

Sugar beet vividly illustrates the power of plant breeding to change plants. In fewer than ten human generations, through the application of selective breeding, the sucrose content of sugar beet, unknown in antiquity, had risen from 5–6 per cent dry weight to approximately 20 per cent dry weight in modern varieties. Sugar beet thrives in temperate climates, in comparison to the tropical climates needed by sugar cane. Furthermore, beets can be stored before sugar processing, whereas sugar cane must be processed within a matter of days otherwise sucrose drops below levels that are economically viable to extract. Consequently, efforts are under way to breed tropical and subtropical sugar beet varieties. More controversially, sugar beet breeding is at the heart of concerns expressed about the release of genetically modified organisms into the environment. Some people are convinced that such organisms are the only way to maintain food production in the face of pests and diseases, climate change and population growth. Others are concerned about the effects such organisms might have on human health; yet others are concerned with the promiscuity of beets and the very mechanisms that led to sugar beet evolution in the first place.

Millennia-old medicinal uses of beets have all but disappeared, except in fringe homeopathic concoctions, although their value as leaf and root vegetables remains. However, it is sugar beet, the newcomer in the beet family, which has become globally significant. Since the 1960s global annual sugar beet production has doubled to approximately 270 million tonnes. Today, sugar beet provides almost one-third of the world's annual sugar production and is an important animal feed. However, sugar beet breeders are also exploring new markets – for example, the supply of raw materials for biofuel production.[37]

The beets have proved highly adaptable to our way of life. Today, with the application of genetic technologies, there is every possibility they will continue to play a major part in our evolution.

Opium poppy

Papaver somniferum L.; Papaveraceae

BOON AND BANE

Plant conservation is often justified because plants are used as medicines, and we do not know what might be important to us in the future. Across Eurasia, in the pre-anaesthetic era when medical treatments were often limited and brutal, the opium poppy was one of the few sources of effective pain relief available.[38] The Italian physician Angelo Sala summarized his opinion of opium in the subtitle to his work *Opiologia* (1614): 'for the comfort and ease of all such persons as are inwardly afflicted with an extreme griefe, or languishing paine, especially such as deprive the body of all naturall rest, and can be cured by no other meanes or Medicine whatsoeuer'.[39] The seventeenth-century English physician Thomas Sydenham argued that 'every choice and noble remedy … received its principal virtues from Nature' and paired *lachryma papaveris* (poppy tears) with quinine as the most 'noble' medicines from the 'hands of God'.[40] Despite such plaudits, throughout our long association with opium we have been, to quote Edgar Allan Poe, 'a bounden slave in the trammels of opium';[41] opium has been a source of immeasurable misery through personal or institutional addiction, corruption and criminality.

The opium poppy is indigenous to Asia Minor and is an annual herb with blue-green, usually hairless, leaves and white sap. Flower colour and fruit shape colour are variable, making it an attractive ornamental plant. When the fruits dry a series of small pores open, dispersing the tiny black seeds in a pepper-pot fashion. But there is negligible morphine (the principal opiate in opium) in poppy seeds, which are widely used as a flavouring and oil source[42] – it is the milk-white sap, transported around the plant by a system of latex vessels, which is rich in morphine and other opiates.

Opium poppy remains have been reported from Neolithic, Bronze Age and Iron Age sites across Europe. Many classical cultures of the Mediterranean, North Africa, Near East and Central Asia used and cultivated poppy.[43] Opium holds an important place in the medical texts of the ancient world – for example, the Egyptian *Ebers Papyrus*, Dioscorides' *De Materia Medica* and Ibn Sina's (Avicenna) *The Canon of Medicine*. The information in many of these manuscripts was copied, translated and reworked into standard European medical works until the mid-seventeenth century.

The mechanics of opium harvesting have probably changed little since antiquity. The surface of green, ripening poppy fruit is repeatedly scratched with a lancet to release droplets of latex that dry to form a sticky, brown resin. With care, 1 hectare of opium poppy can produce up to 12 kilos of raw opium. For the illegal market, raw opium is usually processed into morphine base or heroin for easy smuggling.

Plant use in modern medicine follows a familiar pattern: plant extracts produce relief, an active compound is identified and isolated, and then perhaps artificially synthesized. Until the early twentieth century most medicines were made directly from plant extracts. The advantage of a purified compound is that specific doses can be given: crude plant extracts rely on the apothecary guessing the amount of active ingredient – too little and the drug is ineffective; too much and the patient may die.[44] The trend towards purified drugs began in the early nineteenth century with the isolation of morphine from opium poppy by the German chemist Friedrich Sertürner; a breakthrough that heralded the purification of other plant-based drugs such as nicotine, caffeine and quinine. Commercialization of pure morphine rapidly followed, although morphine use declined as anaesthetics such as chloroform and ether became popular. However, the availability of pure morphine did not make opium resin redundant. Laudanum, that mainstay of

a. Papaver sativum flore saturate pur= purgo pe= talis laciniatis. b. Papaver sativum rubrum flore simplici. c. Papaver sativum unguibus rubris flore pleno, Magen. d. Papaver Spinosum Americanum.

the Victorian medicine cupboard, was an alcoholic extract of opium resin, and continued to be a common acceptable feature of soporifics and patent medicines until opiates started to be controlled across the Western world at the end of the nineteenth century.[45]

When the semi-synthetic opiate heroin was discovered towards the end of the nineteenth century, it was marketed as a non-addictive morphine substitute. During the twentieth century, fully synthetic opiates, including methadone and pethidine, were developed; like all opiates, they are addictive.

Despite the strong Western association of opium use with the Chinese, the recreational use of opium came late to China. Western Europeans appear to have learnt their opium habits from the Ottomans, who supplied opium to Europe long before the Chinese began to do so in the late nineteenth century. Arab traders appear to have introduced opium to China during the first millennium CE but the first reference to Chinese recreational use of the drug was in 1483,[46] and started as a pursuit of the elite, bolstered by beliefs that opium enhanced virility and longevity.

Opium became a major issue of trade policy between imperial Britain and China during the eighteenth and nineteenth centuries. Early in its rule, the Qing dynasty (1644–1912) had opened the port of Canton (modern-day Guangzhou) to foreign traders. The British East India Company started to use Canton to export Chinese tea to feed the growing British obsession (see TEA, page 229). However, since the emperor was only interested in silver, vast amounts of the metal were moving east. Rather than see the depletion of British coffers, the British East India Company used cheap opium grown in British India as a high-value trading commodity. In 1729 the emperor banned the sale of opium, but this hardly curtailed the Company's trade; chests of opium were simply sold in Calcutta and smuggled into China. Over the next century the illegal opium trade became the world's most valuable commodity trading operation and the British government became the world's largest drugs pusher.[47] Encouraged by British imperial policy, opium use became a commonplace pursuit in all walks of Chinese life; one estimate implies that 27 per cent of China's adult male population regularly used opium at the start of the twentieth century.

In 1838 the Chinese destroyed a consignment of opium, to which the British eventually responded by launching the First Opium War; the Chinese lost and conceded Hong Kong as reparation. The Chinese also lost the Second Opium War in 1858 and were forced to legalize opium; they responded by increasing domestic opium production until by the early twentieth century China was supplying 85 per cent of the world's opium.[48] The repercussions of this ignoble period of Sino–British relations resonate in today's war against drugs, where the West suffers at the hands of the drugs pushers.

Opium use, legal or otherwise, is a prominent feature of artistic endeavours; this is not a new phenomenon. Thomas de Quincey's *Confessions of an English Opium-Eater* (1822) presents the pleasure and pain of opium addiction, while laudanum addict Samuel Taylor Coleridge's *Kubla Khan* (1816) apparently describes an opium trip. In European literature, as the Victorian era dragged on and gave way to the Edwardian period, recreational opium use became associated with criminals, foreigners and social outcasts. In Charles Dickens's unfinished *The Mystery of Edwin Drood* (1870–71) we meet Princess Puffer, the hag who runs an opium den visited by the novel's supposed villain. Sax Rohmer's Fu Manchu stories, from the first half of the twentieth century, play to a Chinese stereotype of opium-induced criminality and violence. However, Arthur Conan Doyle's Sherlock Holmes is more ambiguous; opium dens are the haunts of Holmes's opponents but it is acceptable for him to take morphine in his own rooms.

The world is awash with illegally produced opium, as political expediency means governments effectively turn a blind eye or even actively condone its production and use. Yet even at the current level, global opium production is only about a fifth of what it was at the beginning of the twentieth century. While China dominated the opium trade in the 1900s, the economic ebb and flow saw its position usurped in the 1920s by the Golden Triangle that straddles Thailand, Myanmar and Laos. Until late 2023 the main producer was Afghanistan, with 82 per cent of the illegal market, when Myanmar become the world's primary source of illegal opium.[49] In contrast, legal production of opium in India and Turkey cannot keep up with demand. While opium's dark side is what usually makes the headlines, the presence of opium poppy fruits on the Royal College of Anaesthetists' coat of arms is a reminder of its legacy and role in modern medicine.

Brassicas

Brassica oleracea L.; Brassicaceae

OF CABBAGES AND KALES

Cabbages, kale, broccoli, cauliflower, kohlrabi, turnip, Brussels sprouts, horseradish, rocket, mustard: among our domesticated vegetables the brassica family is one of the most diverse. Some are so familiar that they have entered the collective consciousness as rather boring vegetables, and have become social clichés of class, wealth, education and status.[50] But the brassicas provide much more than the greens on our plates, and have played their part in plant breeding, medicine and warfare.

In seventeenth-century England seeds collected from English cabbages were thought inferior to those collected from European cabbages. The Oxford Professor of Botany Robert Morison related the travails of Richard Baal, a gardener from Bramford, Suffolk, who collected and sold seed of what he claimed were premium English cabbages. He was forced to pay compensation when plants raised from the seed proved to be more like wild sorts. With our current knowledge of the cabbage's promiscuity, Baal's woes can be explained – his premium cabbages had crossed with 'inferior' sorts.[51] The diversity of brassicas we see today is all a result of selective breeding at different times by different cultures in different parts of their global range, but all are ultimately from a species of wild cabbage. The ease with which wild cabbage morphology can be changed was demonstrated in 1860, just one year after Darwin published *On the Origin of Species*. Researchers at Cirencester Agricultural College used simple selection experiments to produce broccoli and other cabbage-like forms from wild cabbages found on the English coast.[52]

Braſsica cauliflora, Chou fleur,
Käſe = Kohl.

a. *Braſsica capita-ta, purpurea et alba, Chou cabu.*
Geſtreiffter Kopff-Kohl.
b. *Braſsica capitata cum flore.* Blühender Kopf-Kohl.

Variations in leaf form and terminal bud development produced plants either with loose heads, such as the kales, or the tightly packed heads of today's cabbages. Focusing on thickened stems and various flower formations resulted in cauliflowers and broccolis, while kohlrabi has a grossly enlarged stem, and Brussels sprouts have greatly expanded lateral buds.

When Darwin came to investigate crop origins in the mid-nineteenth century he complained that 'botanists have generally neglected cultivated varieties, as beneath their notice'.[53] The fleshy parts of plants are poorly preserved in archaeological remains, but in the course of the past hundred years or so developments in fields as diverse as physics, chemistry, genetics and computing have all contributed data to the search for the origins of brassicas.

Kales are the most ancient cultivated brassica forms and appear to have developed in the eastern Mediterranean and Asia Minor. Early brassica domestication extends beyond the Mediterranean and Levant, into the southern parts of China, to provide the origin of Chinese kale and the multitude of oriental variations.[54]

Broccolis and cauliflowers were also familiar in the Mediterranean world. However, kales are not well suited to the continental climates of Northern Europe, where cold-tolerant headed cabbages were found to fare better. Headed cabbages were apparently unknown in the ancient world, and poor descriptions make it difficult to know whether or not headed cabbages were known by medieval times. But by 1536 the Frenchman Jean Ruel gives an unmistakable description of a white headed cabbage, and the western migration of headed cabbage started soon after, when the French navigator Jacques Cartier introduced it to Canada in 1541. By the mid-seventeenth century headed cabbage was being cultivated in the North American English colonies.

In the sixteenth century cauliflowers were still rare in England, and their name, Cyprus colewort, emphasized the eastern Mediterranean source of the original seeds, but by the beginning of the next century

Brassica capitata alba, Chou blond, Weißer Kopf-Kohl.

they had become common in London markets. Kohlrabi and Brussels sprouts are much more recent cultivated forms. Kohlrabi was selected in late-fifteenth-century Northern Europe, and the first descriptions of it were made by the Italian botanist Pierandrea Mattioli. By the end of the sixteenth century it was known from the Iberian Peninsula in the west, Libya in the south and across the eastern Mediterranean. Kohlrabi was not grown extensively in England until the late nineteenth century. Sprouting broccoli was not grown until the early eighteenth century, but its popularity has increased hugely in recent years; today the Calabrese type, with its large, green, watery, cauliflower-like heads, is the most frequent broccoli form consumed by westerners. Brussels sprouts are the new kids on the block, as they only became commonplace during the twentieth century.

Many people are surprised to find that the diverse brassica family includes such un-cabbagelike members as turnip, horseradish, rocket and mustard. This last is a clue to their common ancestry: it is the chemicals in mustard oils that give brassicas their distinctive taste. The brassicas evolved these pungent, volatile oils as part of their arsenal against insect attack – and the same oils became a weapon in human hands, deployed devastatingly as mustard gas.

One last member of this varied family remains: oilseed rape. A crop in Europe since medieval times, oilseed rape emerged as a twentieth-century global industrial commodity for oil and fodder. Rape is a hybrid between turnip and cabbage, which appears to have originated on numerous separate occasions.[55] In the early 1960s global oilseed rape production was just under 4 million tonnes; by 2022 production had risen to approximately 87 million tonnes. Furthermore, oilseed rape has become a model organism in plant sciences research and a cause célèbre in modern plant breeding.

Through generations of selection we have changed almost every part of wild cliff-growing cabbages to our varying needs. The evolutionary malleability of cabbage bodes well for its future associations with us as climates change and we must push to ever greater extremes to grow food.

Cannabis

Cannabis sativa L.; Cannabaceae

HEMP'S HIGH LIFE

For thousands of years an intricate cultural tapestry has been woven by the complex interactions among cannabis and the peoples with which it has travelled. We are most familiar with cannabis as marijuana, the drug that helps people encounter their gods, escape their daily lives or, perhaps, heal their ailments. But cannabis is also hemp. Hemp sails and ropes linked outposts of empires, and hempseed oil is both food and lubricant. In China, cannabis has been continuously cultivated, as both fibre and food, for at least 6,000 years, whence its cultivation and use spread globally. In the West, hemp fibre, fashioned into canvas (which derives its name from cannabis), and paper are the surfaces upon which some of the world's great works of art have been wrought.

The consensus that drug and fibre do both come from the same species has only been reached after several centuries of often acrimonious botanical discussion. *Cannabis sativa* is the sole member of the genus *Cannabis*, although two subspecies are recognized, both of which have wild and cultivated forms. In one subspecies, ssp. *sativa*, stems and fruits have been selected for fibre, food and oil production, while in the other subspecies, ssp. *indica*, flowers and leaves have been selected for drug production. The crux of the scientific debate has been how to account for complex patterns of morphological variation found in a widespread species that has been intimately associated with humans for thousands of years. In the early 1970s this arcane academic nomenclatural debate briefly leaked into the public arena when lawyers tried, unsuccessfully, to find clever, semantic loopholes in United States drugs legislation.[56]

Cannabis is a tall, annual, wind-pollinated herb, with separate male and female plants. The hand-like leaves, with their narrow, serrated leaflets, are immediately recognizable to many through its status as a countercultural symbol. Strict legislation surrounding cannabis cultivation means the living plant is not a common sight in fields, although it occasionally pops up in rubbish dumps and in gardens, and the fleeting sweetish odour of cannabis smoke is familiar on city streets.

While the Chinese domesticated cannabis for fibre, inhabitants of the Indian subcontinent focused on its drug properties. The psychoactive drug THC (tetrahydrocannabinol) is concentrated in minute glands that cover the flower buds, especially in female plants. Typically, THC content is below 0.3 per cent dry weight but may reach 20 per cent dry weight in some drug-producing cultivars. Ritualistic use of cannabis, which dates from at least the third millennium BCE, is found through Indian, African, European, Chinese and Central Asian cultures. Cannabis fragments dating back about 2,500 years have been discovered in the grave of a shaman from north-western China.[57] In fifth-century BCE Scythia, Herodotus reported the ritualistic use of smoke from burning cannabis fruits. Cannabis is associated with Sufism and Sikhism, and has become an intimate part of Rastafari culture. As a drug, cannabis hides under hundreds of different, often subculture-specific, names, including bhang, ganja, grass, hash, hashish, skunk, wacky baccy and weed. Cannabis is the world's most popular recreational drug, despite its use and cultivation having been illegal in many Western countries since the early 1940s. Over the last fifty years, numerous small syncretic religious groups have adopted cannabis into their sacraments. Others have made great claims for its medicinal value.

Fibre-producing cannabis contains very little THC. The mechanical rather than transcendental properties attracted the Chinese and most early cannabis growers. In the age of sail, keeping a navy afloat meant having a constant supply of hemp for sails and rigging; 'our whole

Cannabis foemina. 1-3. Frucht 4.5. Saame — Hanf Hanf-Weiblein.

mercantile, as well as royal maritime power, depends on supplying ourselves with cordage' wrote Malachy Postlethwayt, an eighteenth-century trade expert. With a warship needing about 80 tonnes of hemp, equivalent to the output of 400 acres of land, hemp production was a major strategic issue for naval powers and competed with food production.[58] Successive British governments were therefore exercised with the strategic problem of hemp procurement; Elizabeth I even threatened her subjects with large fines if they did not grow it. To stop hemp ropes rotting they had to be regularly tarred; but gradually maritime uses became redundant as alternatives to hemp emerged. Manila hemp, isolated from an East Asian banana, was stronger than hemp and did not need to be tarred. But it came from a restricted, volatile area of the world, and when the Japanese cut off Manila hemp supplies to the United States during the Second World War rope became a strategic issue. The United States government, wary of hemp because of its associations with drugs, was forced to create a brief resurgence of interest in hemp cultivation.

The Bible and bank notes have been printed on hemp-rich paper, while today most cigarettes, including joints, are wrapped in it. Oakum, a traditional tarry naval sealant, was made from hemp recovered by laboriously separating the strands of old ropes; an activity for the poor and incarcerated in Victorian Britain. In Shakespeare's *Henry V*, Pistol pleads for help to save Bardolph's life, alluding to one of hemp's Elizabethan names, neckweed, and iconic uses: 'let not Hemp his windpipe suffocate'. Today, the high costs of growing cannabis and extracting the fibre mean there is little more than a niche market for hemp rope, papers or textiles.

In its long domesticated life, cannabis has moved from being lauded in Western civilization as an economically important fibre to being vilified as a drug. A plant that allowed Western nations to control vast areas of the globe, and then considered so dangerous that its use might tear them apart, is now becoming legalized by some.

Bread wheat

Triticum aestivum L.; Poaceae

CHANGING PEOPLE'S SHAPE

Wheat, the staff of life. It is the most important food grass in Western Europe and has, arguably, sustained the development of Western civilization, philosophy, religion and science. The grain, rich in starch and protein, is a concentrated, compact energy reserve, which enabled societies as diverse as those of Sumeria, Egypt, Greece and Rome to develop and transform the physical and intellectual ancient world.

Wheat belongs to a small genus of annual grasses whose distribution is concentrated in the Fertile Crescent of the Near East.[59] Wheat fragments from archaeological sites show that the Fertile Crescent was the cradle of Western agriculture. The wild wheats, and their close relatives the goat grasses, produce grains annually, grow in large populations through the region and are readily harvested. Consequently, in the past, nomadic or semi-nomadic peoples could have had access to predictable food supplies. Gradually, humans started to collect and farm the grains, and we became sedentary. And we also became changed by the plants we farmed.

The centuries-old story of how the grains gathered by our ancestors changed into the fat wheat ears we know today is a good example of evolution in action. Between a half a million and 3 million years ago, wild wheat naturally hybridized with a goat grass to produce another type of wheat, wild emmer. Wild emmer was a successful evolutionary experiment; it spread throughout the Near East long before its use by humans. In the Jordan Valley and Syria, early hunter–gatherers 23,000 years ago were using wild emmer and another type of wild wheat, einkorn, and about 10,000 years later domestication of both wild

I.II.III. *Triticum Spica mutica*
IV.V. *Triticum ariſtis longis.*

2.-21. Blume.
22.-28. Frucht.
29.-36. Saamen.

Weitzen
Grät oder Barth-Weitze

emmer and einkorn started. Cultivated emmer, together with barley, became the staple of civilizations stretching from Egypt through Mesopotamia to the Indus Valley. Some seven to ten millennia ago, in Transcaucasia, hybridization between cultivated emmer and another goatgrass produced what we now know as bread wheat.[60]

Wheat domestication is the product of astonishing artificial modification of natural genetic variation. Generations of hunter–gatherers collected the wild wheats for food without changing the plants substantially. However, when hunter–gatherers engaged in the apparently simple act of picking a grain of wild wheat, saving it and then planting it, they had profound effects on the plant and on themselves. To be successfully domesticated, wheat had to be a good food and have the right genetics. Adaptations that eased harvesting and increased grain production were important socially and biologically. Importantly, they also provide archaeological markers to understand cultures across Western Europe through the Mediterranean, North Africa and the Levant to Central Asia and northern India. The most conspicuous harvesting feature is that mature grains in cultivated wheats remain attached to the plant, while wild wheats drop their grains at maturity. Theoretically, wheats that retain their grains can be selected from those that drop their grains in as little as twenty years, although archaeological evidence shows this transformation took around 1,500 years.[61] About 2,000 years ago bread wheat production overtook emmer production.

For much of its history bread wheat has had social cachet. It can be used to make light, easily digested bread, but, since the species' geographic range was limited, it was expensive. The rich flaunted it, the poor desired it, and if it could be said that bread wheat was an emblematic marker between wealth and poverty, then, come the revolution, monarchs lost their heads over it.

A wheat field, even a very basic field, is a very different environment from the wild habitat. Simply scratching the land with a primitive plough and weeding would alter the growing conditions; competition among species would be reduced but competition between cereal seedlings would be all the more intense. Those grains that germinated first or produced the biggest seedlings would produce the plants that would be harvested for the next generation. As a result cultivated wheats have large, uniform, short, fat grains, while wild wheats have small, variable, long, thin grains.

It is easy to see how wheat became socially, economically and politically important, but it has also changed us genetically. Cereal-based agriculture increased the quantity and predictability of the food supply, but not necessarily its quality. Compared to the diets of their hunter–gatherer ancestors, early agriculturalists consumed few proteins and vitamins but a lot of carbohydrate. Humans had spent generations genetically adapting to hunter–gatherer lifestyles, so were, initially, probably poorly adapted to life as sedentary farmers.[62]

Wheat grains contain compounds detrimental to health; for example, they have high levels of compounds that interfere with iron absorption. Consequently, skeletons of people raised on wheat-rich diets show symptoms of iron deficiency; the more intensive the agriculture, the more pronounced the symptoms.[63] The archaeological record also shows people in agricultural societies suffered from rickets (vitamin D deficiency). We get vitamin D from our food or via ultraviolet light (sunlight). African hominids evolved deeply pigmented skin to protect them from harmful ultraviolet radiation, but as their descendants migrated out into less sunny areas rickets become prevalent. Northern populations adapted in two ways.[64] Skin tones lightened, and, unusually, adult Northern Europeans started to consume milk. The lactose-tolerance mutation appears to have originated about 8,000 years ago and spread rapidly though Northern Europeans. Ötzi, the 5,300-year-old Iceman whose corpse was discovered in a glacier in the Italian Alps in 1991, was lactose-intolerant.[65] Although the adoption of agriculture correlates with a dramatic decline in individual health, some members of society benefited. The lives of hunter–gatherer women were shorter than the lives of their menfolk; they had to endure the gamble of childbirth and the stresses of child rearing. However, in the eastern Mediterranean, following the adoption of agriculture, the lifespan of women increased – a farming life appears to have been a little less risky for women.

Wheat influenced not only our digestive systems and state of health, but our size and shape. Cereals are easier to chew than wild-collected plant foods. Consequently, the agricultural life led to changes in the human face and teeth; jaws have become more delicate and teeth smaller. More dramatic still was the effect of cereal farming on adult height. In the Near East, coincident with the spread of extensive agriculture, there was a steady decline in adult height until about 4,000 years ago; and even today we remain about 3 cm shorter than our hunter–gatherer ancestors.[66]

In Western cultures wheat has remained our cereal of choice, whether in bread, cakes or pasta. Millers, the commodity traders of their day, were wealthy societal hate figures. Chaucer satirized a popular view in his fourteenth-century *The Canterbury Tales* when the reeve tells a tale of a dishonest miller cuckolded and humiliated. As wheat has become an international commodity, prices have become subject to speculators more interested in profit than food supply. Wheat breeding, along with the scientific culture it engenders, has become essential for future food security. If we cannot feed ourselves, our future prospects are limited.

Over thousands of years the palette of wheat genetics has been filled. To continue the analogy, we have started to mix our colours to produce the wheat plants best suited to our purposes. At first mixing was blind, driven by luck more than judgement, but as we have understood more our mixing became more objective. The vision of many, and the nightmare of some, is that we will start creating our own 'colours', transferring genes between distantly related species, rather than relying on natural variation among closely related species, to create the wheat varieties needed for the future.

Faba maior.

Grose Bohnen.

Broad bean

Vicia faba L.; Fabaceae

FOOD, FODDER AND FERTILIZER

Before Columbus's discovery of the Americas, beans, to the peoples of Europe, the Near East and North Africa, meant broad beans. Broad beans were among a quartet of legumes (along with peas, chickpeas and lentils) growing in gardens and fields, to which the economies of multiple Western civilizations – if you'll forgive the pun – pulsed.

Despite their ubiquity, beans were not universally loved. While Theophrastus, a follower of Aristotle and the father of botany, discussed in detail the biology and cultivation of beans, and emphasized their value as food, a fellow Greek, the mathematician Pythagoras, was of a markedly different view. He banned his disciples from eating beans and, at the end of his life, apparently refused to cross a field of flowering beans to escape being killed. This has led to speculation he suffered from favism, a hereditary blood condition that is brought on by exposure to specific alkaloids in broad beans.[67] Others have speculated the reasons were more superstitious: immature beans resembling 'privy parts' could lead to lewd thoughts, or exposure of beans to the moon turn them into blood.[68] The Ancient Egyptian priesthood also considered beans taboo.[69]

The medieval Latin proverb 'where beans thrive, fools grow' plays to Ancient Greek and Roman notions that the heady scent of broad bean flowers induces nightmares and madness.[70] Despite the olfactory dangers, broad beans have been grown across the warm, dry Mediterranean and cooler temperate regions of Europe and Asia for thousands of years, both as a protein-rich food for humans and their livestock and as a nitrogen-rich green manure. The Roman agriculturalist Varro pointed out that 'some crops are also to be planted not so much for the immediate return ... it is customary to plough under ... field beans before the pods have formed ... in place of dung, if the soil is rather thin.'[71] Varro was referring to the ability of leguminous plants such as peas and beans to store nitrogen in a form usable by other plants, and it is notable that practical agriculturalists were aware of and taking advantage of this valuable facility more than 2,000 years ago (see also PEA, page 65).

As well as being a common staple of the dinner table, broad beans were important in Roman religious, social and political life.[72] At the ceremony of Fabaria, associated with the god Janus, offerings of broad bean meal were made, while dried beans were considered lucky in agriculture and trade. Votes were cast using white and black broad beans; this survives today in the discrimination of being blackballed. Broad bean flowers are butterfly-like, large and white streaked and blotched with purple, and some people divined images among the spots and lines, giving the plant a sacred status. The belief that human souls resided in broad beans meant they were offered at funerals. One of the great Roman dynasties, the Fabii, was named after the beans they once cultivated. Romans eased away signs of ageing using cosmetics containing bean meal, and they also used beans as medicine and fish bait, and even made use of beanstalk ashes in glass-making.

Beans were thought by many to cause flatulence, and one can but imagine how broad beans added to the odours emitted by crowded communities whose livelihoods depended upon them. Beans, as 'flatulent and coarse food', were also associated with social status, since they were more 'suited to the laborious, than the sedentary class of society'.[73] Furthermore, in the eighteenth century field beans were loaded onto slave ships leaving Bristol in readiness to be fed to their human cargoes during the 'middle passage' of the slave triangle across the Atlantic.[74]

By the beginning of the nineteenth century, in Britain beans were 'seldom, if ever, used as food, in this improved country, in their dried state; but when sent to table young, they [were] generally admired'.[75] Garden beans and field (or horse) beans were thought to be different species, although gardeners such as the sixteenth-century herbalist John Gerard were not fooled; he believed the differences between the two were due to the richer soils and additional care bean plants got in the garden.[76] Many different cultivars of broad beans were selected across Europe, but one of the best, the Mazagan, was originally collected from a Portuguese settlement on the Moroccan coast, although at least one nineteenth-century vegetable breeder thought it 'the worst and most useless of its race'.[77]

Broad beans' botanical origin remains an enigma, despite the technological and intellectual advances made over the last century in understanding crops and their wild relatives. The overall appearance of the plant – a rather rugged, upright, non-climbing annual with large blue-green hairless leaves – suggests its closest wild relatives are a group of Mediterranean and Near Eastern vetches. If this were the case, crossing these wild vetches with broad beans should produce hybrids, yet decades of research have failed to make such hybrids. Clues as to why the hybridizations fail have come from studying the plants' chromosomes. Cultivated plants and their wild relatives are expected to have identical chromosome numbers, or whole multiples of these chromosomes. However, broad beans have 12 chromosomes, while the wild vetches have 14 chromosomes. Not only are broad beans reproductively and cytologically unique; detailed DNA analyses reveal they are genetically distinct from all other *Vicia* species.[78] These data imply that either the wild relative is as yet undiscovered – meaning it is rare and has a restricted distribution – or it is extinct. Or, as broad beans have never been found outside cultivation, it may imply they have originated through domestication.

Remains of legumes similar to broad beans are common at archaeological sites across the Mediterranean and Near East. However, broad-bean-like vetches are also common in the region, which means precise identification of broad beans is difficult. These problems become more acute in early archaeological contexts, and where there are few beans recovered or the beans are charred. Consequently, early reports identifying charred seeds from Jordanian (11,000 to 13,000 years old) and Syrian (*c.* 10,000 years old) sites as broad beans are controversial.[79] The earliest undisputed broad bean remains come from a site about 8,500 years old near Nazareth, Israel. Other major archaeological sites for broad bean remains are found in the countries bordering the Aegean (4000–2000 BCE), northern Italy and Switzerland (4000–1000 BCE) and the Iberian Peninsula (3000–2000 BCE).

Beans played three roles in pre-twentieth-century Western civilizations: as human food, as animal fodder and as a nitrogen fertilizer. However, industrially synthesized nitrogen fertilizers and other food and fodder stuffs have become commonplace. Global broad bean production gradually declined from the early 1960s, until about a decade ago, when production once more began to increase. In 2022 global production was nearly 8 million tonnes.

Alliums

Allium species; Amaryllidaceae

SIGHT FOR SORE EYES

In Gabriel Axel's film *Babette's Feast* (1987) the eponymous heroine enlivens the bland diet of a tiny Danish religious sect by introducing onions into their food. Onions and their cousins in the allium family, including leeks, garlic and chives, may not be essential for human existence, but their flavours do make life more interesting.

The characteristic sulphurous odours of alliums may also be responsible for a Near Eastern superstition that onion and garlic sprang from Satan's right and left footprints respectively, as he emerged from the Garden of Eden.[80] Despite such diabolical stories, alliums are consumed across the globe and have become imbued with all manner of marvellous properties. In the seventeenth century garlic was a panacea apparently curing everything from 'obstructions or stoppings' through the bites of 'venomous creatures' (such as dogs and shrews) to dandruff, toothache and stomach ache.[81] And, of course, it was well known for its power to ward off vampires. Today, the benefits attributed to garlic are different, but its reputation has grown; it is claimed to have anti-inflammatory, antibiotic, anti-cancer and antioxidant properties, although supporting data relevant to such claims are frequently contradictory.

Alliums were among the first plants recorded by the civilizations that have contributed to Western cultures. In Mesopotamia, six millennia ago, instructions for the cultivation and use of alliums were baked into cuneiform bricks, while Ancient Egyptian records are full of allium references;[82] accounts of onion growing are recorded in the tombs of the Old Kingdom. The fifth-century BCE Greek historian

Herodotus recorded that an inscription on the Great Pyramid of Cheops (*c.* 3700 BCE) reported that the Egyptian state paid 1,600 talents of silver for the supply of vegetables, including onions and garlic, to the workers who built the edifice.[83] Despite the importance of garlic as a food in Ancient Egypt, the priesthood avoided it.[84] When the emancipated Hebrew slaves were wandering through the Sinai Desert, in search of the Promised Land, they yearned for their Egyptian comfort foods:

> We remember the fish, which we did eat in Egypt freely; the cucumbers, and the melons, and the leeks, and the onions, and the garlick. (Numbers 11:5)

To this day, alliums remain comfort-food ingredients.

Being water-rich, alliums are unlikely to be preserved in archaeological deposits and yet, under the right environmental conditions, they do survive. Excellent allium remains have been found with preserved bodies of Egyptians from the Eighteenth Dynasty (fifteenth–twelfth centuries BCE), including entire garlic bulbs from the tomb of Tutankhamun,[85] and Israel has yielded the earliest unequivocal physical evidence of garlic use, in deposits some 6,000 years old. Despite the general absence of archaeobotanical evidence from outside Egypt and the Near East, alliums must have been used for millennia across Eurasia.

The genus *Allium* comprises more than 700 species, distributed across the Old and New Worlds. Wild alliums were probably used wherever humans found them. As Eurasians colonized new lands, their foods and their ideas followed. Just as indigenous cultures gave way to the colonizers, native alliums gave way to leek, onion and garlic. Different environments and different tastes, together with thousands of years of cultivation and selection, have produced hundreds of variations across the globe. Garlic was so popular in the classical world that it has even been suggested that the expansion of the Roman Empire

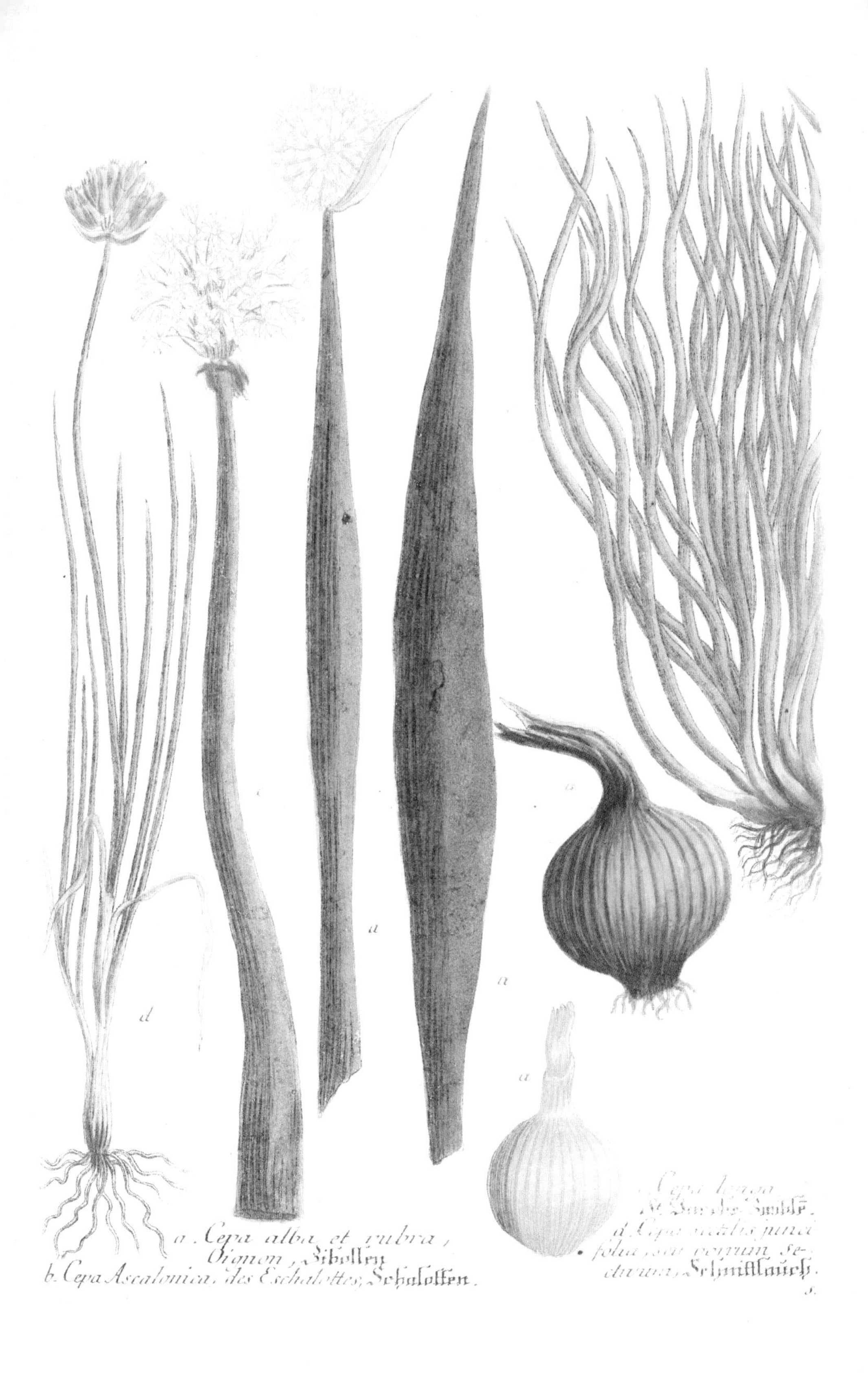
a. Cepa alba et rubra,
Oignon, Zibollen
b. Cepa Ascalonica, des Eschalottes, Schalotten.
d. Cepa sectilis junci
folia, seu porrum se-
ctivum, Schnittlauch.

could be reconstructed by tracing the timing and distribution of garlic cultivation.[86] True garlic and onion are known only in cultivation, as Gerard noted in his *Herball*: 'divers sorts of Onions, which have their surnames of the places where they grow. Some also lesser, others greater: some be round, and divers other long: but none wilde as Pliny writeth.'[87]

The close association of garlic and onion with humans has meant their wild parentages are controversial, with authorities on the genus making the case for different wild species having given rise to the familiar domesticated forms.[88] However, authors tend to agree that wild relatives of garlic and onion probably came from south-western or Central Asia.[89] Leeks, on the other hand, appear to have been selected from a single species of wild leek, which is distributed throughout the Mediterranean region.

Alliums complete their life cycles in two years; hence they need food reserves to see them through the winter months and enable them to produce flowers in the second year. We take advantage of this by usually harvesting alliums at the end of the first year, to use these food reserves – packed into bundles of leaf sheaths in the case of leeks, or in the swollen bulbs of garlics and onions.

The most distinctive feature all alliums share is an unmistakable odour, albeit varying in strength. This is part of the plants' biochemical defence against pathogens and herbivores – damage an allium cell and it will release a complex chemical deterrent. In particular, it releases an enzyme called allinase, which catalyses production of a short-lived sulphenic acid. In garlic, this acid forms allicin. In onions, however, a second enzyme reacts with the sulphenic acid to produce lachrymatory factor – which is why we cry when we chop onions but not garlic. As our understanding of the biochemistry and genetics of these reactions has increased in the past decade, it has been possible to manipulate biochemical pathways to vary the levels of sulphurous precursors and hence vary the flavours of different cultivars; it has also enabled breeders to produce so-called 'tearless onions'.[90]

Cooks have learnt to manipulate this sulphurous chemical cocktail by trial and error. For example, the amount of allicin produced by garlic can be increased by crushing cloves and leaving them for short periods, so the allinase enzyme has time to act and develop the flavour. Cooking temperatures also alter the chemical composition, which is why, for example, stir-fried garlic smells and tastes different from slow-cooked garlic, and leeks or onions processed by drying, freezing or pickling will have different flavours from their fresh counterparts.

In 2022 some 29 million tonnes and 115 million tonnes of garlic and onion respectively were produced globally, plus a more modest 2.1 million tonnes of leeks – alliums remain as important in our diets today as they did thousands of years ago.[91]

a. Pisum majus flore purpureo, Kieff-Erbis.
b. Pisum marinum perpetuum.

Pea

Pisum sativum L.; Fabaceae

THE POWER OF A HUMBLE LEGUME

So familiar is the little garden pea that it is easy to overlook its significance in the history of agriculture and genetics.

Nowadays, the peas we mostly eat are unripe green seeds or even, as mange tout, whole immature pods. This is all thanks to the improvements in picking, canning, chilling and freezing technologies. But for much of our history, right back into antiquity, we have eaten dried mature pea seeds. Peas have frequently been found, alongside barley and wheat remains, at Near Eastern archaeological sites associated with the development of early agriculture, and peas were an important element of the mixture of crops grown as Neolithic agriculture spread through Europe more than 8,000 years ago. Peas had two important functions in these societies: as a protein-rich food for people and as a nutrient source for other crops. This second role, as a fertilizer, cannot be overestimated.

Plants cannot survive on water, fresh air and light alone; they need inorganic nutrients, such as magnesium, potassium and nitrogen. Since nitrogen comprises about four-fifths of the air, it would be natural to assume there would be no shortage, but for plants to gain access to nitrogen they need it to be locked up in the form of soluble salts, nitrates. And in this lies the value of peas and other legumes: bacteria, called rhizobia, live inside nodules attached to legume roots, and these are able to fix nitrogen in a form that plants can use.

The influence of a soil's richness or infertility on crops has long been recognized. Greco-Roman writers, such as Theophrastus, Columella, Varro and Cato, wrote about the key role of soil fertility in agricultural productivity: Varro, the Roman agriculturalist, in the first century BCE, recommended fields be fertilized with dung of all descriptions,[92] and in the *Georgics* the Roman author Virgil commended 'the genius of soils, the strength of each, its hue, its native power for bearing'.[93] Fertile soils are highly complex living mixtures of minerals, organic matter, air, water and microbes. A fertile soil holds moisture, allows easy root penetration and has the microbes needed to break down and release nutrients essential for plant growth. As crops grow, they deplete soil nutrients; consequently, replenishment is essential. In different epochs, successful farmers learnt to modify soil fertility by harnessing the magic of muck and the power of legumes.

In pre-Enclosure Britain, before the mid-eighteenth century, peas and other legumes were important in the crop-rotation systems that maintained soil fertility and kept the British population fed. In addition, farmers and gardeners promoted all manner of nostrums to improve soil quality: the market gardens that fed London were themselves fed on a rich confection of metropolitan ordure. By the nineteenth century other sources of fertilizer were being pressed into keeping Britain's farmland in good heart: bonemeal, manufactured from vast quantities of animal bones imported from European slaughterhouses and the bones of soldiers killed on European battlefields, and bird excrement, following the discovery of enormous South American guano deposits in the early nineteenth century. When in the twentieth century the Haber–Bosch process enabled industrial-scale fixing of nitrogen via chemical processing (see SMOOTH MEADOW GRASS, page 199), peas and legumes became less important as a green manure in Western farming systems. In the kitchen and in the field of genetic research, however, they still had much to offer.

There are two primary types of wild pea in the *Pisum sativum* complex: a tall form found in maquis scrubland throughout the Mediterranean and a short form found in the dry steppe grasslands of Southwest Asia. All available evidence shows the short steppe type was the progenitor of the garden pea. Detailed genetic investigations have shown that at least fifteen genes are critical in pea domestication.[94] This suite of genes controls the 'domestication syndrome', which includes pod splitting (one of the most obvious features of domesticated pea pods is that they do not split and spill their peas), seed size and dormancy, seed coat form and habit. Pea remains up to ten millennia old are frequently recovered from archaeological sites in the Fertile Crescent and the Near East. However, deciding whether they are from wild-collected peas or domesticated peas can be difficult.

In the nineteenth century the French horticulturalist Vilmorin reported more than 150 named garden pea cultivars.[95] Charles Darwin, investigating peas as part of his studies of evolution and artificial selection, grew forty-one English and French pea varieties. The varieties he raised, which bore evocative names such as 'Blue Prussian', 'Bishop's Long Pod' and 'Woodford's Green Marrow', showed remarkable variability in flower, pod and seed characteristics and, as he recorded, ranged in height from '6 and 12 inches to 8 feet'.[96] Today, the need for rapid mechanical harvesting from field-grown peas has changed the types of peas we grow. Once common cultivars have become rare; once rare cultivars have become commonplace.

Peas have proved key to the great intellectual and cultural revolution of genetics in our own time. Modern genetics had its origins in the experiments carried out on garden peas in the mid-nineteenth century by the Moravian monk Gregor Mendel.[97] Although Mendel is rightly lauded as 'the father of genetics', his work was foreshadowed by an English fruit and vegetable breeder, Thomas Andrew Knight. In 1787 Knight started to investigate how plants' characters were passed from one generation to the next. He was pragmatic enough to realize that if he wanted quick results he needed a plant with a short life cycle; the plant he chose was the garden pea. Through hybridizations between different types of pea, Knight made considerable progress in showing how variations were inherited from one generation to the next. However, unlike Mendel more than seventy years later, Knight did not count the different offspring that resulted from the crosses;[98] he appeared more interested in the products of the hybridization than in determining the rules of inheritance that Mendel eventually described. Recognition of Mendel's revolutionary deductions was slow in coming – despite having a copy of Mendel's seminal 1866 paper, Darwin failed to grasp their significance for his own ideas, for example. This insight would only come at the start of the twentieth century, as the new science of genetics was born. And so it is through the common garden pea that we have come to understand the mechanics of how features are passed between generations and how we can manipulate these features predictably, contributing to immeasurable advances in medicine and agriculture.

Olive

Olea europaea L.; Oleaceae

DIVINE OIL

Some plants are so strongly associated with particular peoples that they almost become cultural stereotypes. In the case of the olive, an entire region of Europe and Western Asia is brought to mind. Furthermore, the rugged beauty of an ancient, productive olive tree, the tangy taste of the fruit or the green-tinged golden fruit oil give us contact with ways of life that have all but disappeared.

In the 1780s John Sibthorp, Professor of Botany at Oxford University and botanical explorer of the eastern Mediterranean, waxed eloquent to his students about the olive:

> to the Eastern nations the Olive is an Object of the greatest Importance. The Athenians consecrate it to the Guardian of their Citadel Minerva & they still cultivate it in their Plains ... the Olive in its wild State abounds throughout the East. Vast Woods of it occur on the Road from Ephesus to Smyrna & [on] the Greek Islands we find it growing on the most inaccessible Rocks.[99]

Sibthorp's classically educated students should have been familiar with the wealth of literary references to olives, but their direct experience of olives may have been limited to the medicinal uses of the oil. When demonstrating the plant growing in the Oxford Botanic Garden, Sibthorp felt compelled to state that 'our Olive Tree tho' it produces its flowers, seldom ripens its Fruits & that you may have a more perfect Idea of it I shall show You the fructification in a drawing.'[100]

The olive is one of the world's oldest cultivated plants. Olives have been an economic backbone and article of international commerce for peoples around the Mediterranean and stretching across to western Iran for at least 7,000 years.[101] That the name for the tree and its fruit is bound up in so many similar words across European languages – olive, oliva, oliba, olijf, ulliri – is an indicator of its antiquity; 'olive' derives from Latin through a Greek root, and also gives us the word 'oil' and its derivatives. Greek and Roman myths are full of references to olives and oil trees, while the Bible identifies the olive as one of the plants associated with the Promised Land.[102] Olive branches have become symbols of peace through association with the leaf the dove brought back in the tale of Noah's Ark. Olive oil was part of the rites of classical civilizations, and is still used in Judeo-Christian religious practices. However, olives and olive oil are more than elements of literature and religious ritual; they were, and are, vital food and fuel sources for Western cultures. Tales from *One Thousand and One Nights* are replete with Arabian marketplaces filled with olive traders. Jars of olive oil were recovered from the tomb of Tutankhamun, and olive stone remains, along with dates, grains and pulses, are commonly recovered from archaeological sites across the Near East.[103] The economic importance of olives to the countries around the Mediterranean continues: in 2022 more than 21 million tonnes of olives were harvested.

Olives are medium-sized evergreen trees, with often characteristically gnarled, twisted, squat trunks. The elliptical leaves arranged opposite each other along the young stems are covered in tiny hairs, mushroom-like under magnification, that give them a characteristic silvery green appearance. The tiny white flowers form in elongated clusters in the leaf axils and eventually produce small oil-rich plum-like fruits; olives are harvested when they are green or purple. While wild olives generally reproduce sexually by seed, domesticated olives are usually reproduced clonally – it has been found that domesticated olives raised from seed usually produce poor-quality fruit.

As a rule of thumb, the region of natural olive growth defines the European Mediterranean region. Cultivated olive trees grow through

Olea sylvestris
Oleaster

Wilder Öhl-
Baum.

much of Iberia and in a fringe along the northern and eastern coasts of the Mediterranean Sea. Olive trees occur in pockets on the North African coast, while a tongue of tree growth extends through Turkey as far west as Iran. By the end of the sixteenth century olives had been established in the New World,[104] especially in those regions with warm, wet winters and hot, dry summers, such as California and central Chile. Over the past century, with the diaspora of Mediterranean populations and increased global demand for olive products, cultivation has extended to all the world's Mediterranean regions, including parts of South Africa and New Zealand.

For centuries, Northern European gardeners have tried to capture something of the Mediterranean by cosseting olive trees and myrtles in glasshouses and gardens. In Britain, young olive trees have become something of a cliché of the 'Mediterranean-style' garden, promoted by suppliers as possible garden plants if the country's climate warms in the future. But in a habitat that suits them, olive trees are tough: they evolved in environments where they are subject to drought and fire. Furthermore, they may readily become feral, making them a serious weed risk, especially in regions of the world where they have been introduced rather than spread naturally.[105]

The cultivated olive is a part of a large, variable complex of wild olives.[106] The details of olive domestication and its close wild relatives are poorly known but the available evidence suggests olives were first brought into cultivation in the Levant. Since they were first domesticated, olives have escaped back into the wild, confusing boundaries, while the ebb and flow of olive genes among wild, feral and cultivated trees has created enormous diversity,[107] the tangle further complicated by thousands of recorded olive cultivars. The wild olive of the Mediterranean, with its small leaves, spiny juvenile branches and thin-fleshed fruits, similar to the Niçoise olive, is probably a feral sort of the cultivated olive. When St Paul spoke allegorically of the olive tree, he knew his audience would understand the practical subtleties of olive cultivation and the distinctions between wild and cultivated olives:

> For if thou wert cut out of the olive tree which is wild by nature and wert graffed contrary to nature into a good olive tree: how much shall these, which be the natural branches be graffed into their own olive tree? (Romans 11:24)

Olive trees can be ancient, with reports, often unsubstantiated except by tradition, of individual trees several thousand years old. The gap of scientific doubt around age estimates is usually large enough for the camel of romance to pass through, but ancient trees often remain revered, even if shown scientifically to be much younger than claimed.

Grape

Vitis vinifera L.; Vitaceae

IN VINO VERITAS

If prostitution is the world's oldest profession, then winemaking is probably a close second. The raw material for wine production, the grape, was domesticated from the Eurasian wild vines some 6,000–8,000 years ago in the Near East,[108] and grapes are intimately associated with the natural history of alcohol, and our history of drunkenness. The grape is the first specifically cultivated plant mentioned in the Old Testament: Noah plants a vineyard and, in the following verse, imbibes wine to excess and collapses, naked, in a drunken stupor. Although grapes are popular as fresh or dried fruits, it is wine that consumes the vast majority of the output from the world's 73,000 sq km of vineyards.[109]

Literal and symbolic references to the vine and wine are legion through the arts of ancient and modern Western cultures, especially around the Mediterranean. The uninhibited rites associated with worship of the Greek and Roman gods of wine, Dionysus and Bacchus respectively, give us the adjectives 'dionysian' and 'bacchanalian' and their associations with frenzied and orgiastic behaviour. With their classical, religious and social associations, grapevines were essential elements in sophisticated European gardens from the medieval period, not only as food and drink plants but also as shade and as symbols.[110] Something of the position of the vine in the sixteenth-century garden can be seen in the title page woodcut of Jean Ruel's *De natura stirpium libri tres* (1536) with its riot of plants and animals overarched by a vine-covered arbour.

Vitis Vinifera fructu rubro, Vin rouge, Weinreben, oder rother wein.

The natural range of the Eurasian wild grape, where it scrambles along riversides and clambers over forest trees, extends from Portugal in the west to Turkmenistan in the east and from Germany in the north to Tunisia in the south. However, the wild grape underwent dramatic changes as it was domesticated. The size and sugar content of individual fruits increased, as did the number of fruits per bunch. Wild grapes have separate male and female plants, meaning that only about half the plants in a population will be female, so selecting out hermaphrodite plants meant that fruit production became more regular. The function of the dramatic changes in seed structure associated with domestication is not understood, but they are important since it means wild and domesticated grape seeds may be readily separated in archaeological excavations.

Besides their seeds, grapes provide another archaeological marker, tartaric acid, which is commonly found on vessels used for wine fermentation.[111] Grapes appear initially to have been transported long distances by seed but over shorter distances by cuttings. Their cultivation gradually spread through the Mediterranean region along the trade routes of the Assyrians, Phoenicians, Greeks and Etruscans.[112] The Romans pushed the limits of grape cultivation into the temperate regions of their European empire, even up into the barbaric British Isles. With the fall of Imperial Rome, the new Roman Empire, in the guise of the Catholic Church, and its monastic foot soldiers, kept viticulture alive during the Middle Ages and spread it still further. For the Church, wine was thought important for the health of the body and the soul, since it was used in the Eucharist. In contrast, the cultivation of table grapes through the Middle East, North Africa and the Iberian Peninsula appears to have been associated with the spread of Islam.

In the eleventh and twelfth centuries, even in an English climate, grapes produced passable wine; the *Domesday Book* (completed 1086) mentions thirty-eight vineyards.[113] As the British climate deteriorated during the mini Ice Age, vines still grew but their fruit did not ripen.[114]

Unripe grapes were used to make highly acidic verjuice, an essential ingredient of sophisticated cooking of the period, but wine production became impossible and vineyards gave way to orchards.

Rediscovery of the New World provided the Church with spiritual fodder and climates amenable to grape cultivation. When Leif Erikson discovered the Americas some five centuries before Columbus he called the area he discovered Vinland (Wine Land), a reference to the large numbers of wild grapes his expedition discovered,[115] although indigenous North Americans appear not to have made wine.[116] The importance of the American wild grape was to come to the fore in the 1830s, however, when European vineyards found themselves under attack by phylloxera, a sap-sucking bug introduced from North America.[117] Even faced with potential ruin, it took years of argument before the mostly highly conservative winegrowers agreed that the only effective means of dealing with the devastating threat posed by phylloxera was to graft familiar European grapes onto rootstocks of American grapes that were resistant to the pest.

A wine's so-called *terroir* is an amalgamation of grape variety, the conditions under which it was grown and the yeasts and bacteria used in its fermentation.[118] The effect of the yeast is to convert the fruit sugars of grape juice, glucose and fructose into ethanol and carbon dioxide – it is captured carbon dioxide that provides the bubbles in champagne. Yeast and bacteria also produce complex aromatic compounds, esters and organic acids, and a special type of wine bacteria produces chemicals that give some wines their caramel flavours. Around such details wine connoisseurs have developed an exclusive language. The rich array of flavours and odours developed in a bottle of wine may send not only the taste buds of some connoisseurs into overdrive, but also their prose.

Today, there are up to 10,000 different grape cultivars known.[119] The majority are found only in specialist collections; a tiny handful are grown commercially, such as Pinot Noir, Cabernet Sauvignon,

Chardonnay and Thompson Seedless, and these cultivars are often very closely related.[120] Pinot Noir is thought to date back to the first century CE, with Chardonnay being traced to the fourteenth century as a cross between Pinot Noir and Gouais Blanc. Cabernet Sauvignon, the product of a cross between Cabernet Franc and Sauvignon Blanc, appears to have emerged in the early eighteenth century, while thousands of white grape cultivars can be traced to serendipitous mutations in two genes of one red grape cultivar.

Grape production is not a genteel cottage industry: it is a constant battle to keep the vines under control and free from pests and diseases. Their close genetic relationship, as well as propagation techniques, such as grafting and cuttings, mean vines are poorly adapted to resist the battalions of rapidly evolving fungal pathogens and insect pests that bombard them. Consequently, growers use arsenals of ever more sophisticated agrochemicals, despite the financial and environmental costs.[121] Breeders, engaged in the laborious, expensive search to augment grape variation, are sometimes hamstrung by the desire of wine producers and consumers for familiar, traditional cultivars.

If wine drinkers are going to continue to enjoy 'bottled sunlight', they will have little choice but to adapt, as they have in the past. In the nineteenth century they adapted to the invasion of phylloxera; in the twentieth century they adapted to the threat posed by the shift of wine production centres to the Americas, Australasia and South Africa. And now, over the next century, it is likely the commercial demands of viticulture will bring them hard up against the Scylla of conservatism and tradition and the Charybdis of changing climates and plant-breeding technologies.

Papyrus

Cyperus papyrus L.; Cyperaceae

FOR THE RECORD

Libraries and archives are cultural crossroads of knowledge exchange, where the past transmits information to the present, and where the present has the opportunity to engage the future. Bureaucracies become the backbone of civilizations, as governments try to keep track of populations, business transactions and taxes. At a personal level, our lives are governed by the documents we possess; we are certified on paper literally from birth to death. And written documentation carries enormous cultural importance: consider the consequences of signing the Treaty of Tordesillas, the United States Constitution, the Munich Agreement, the Foundation Document of the United Nations or the Convention on Biological Diversity.

Documentation requires some form of alphabet to record information and a surface upon which to record that information permanently. Sumerians recorded their deeds in cuneiform on baked mud blocks, but these, despite being fireproof, were difficult to store. Other cultures recorded information on more flexible but less permanent surfaces, including animal skins, wood strips and palm fronds. In Western culture the adoption of papyrus was to have a great impact. Papyri not only provide an invaluable record of people's daily lives; they can be dated – using carbon-dating techniques it is possible to determine when the plant used to make particular pieces of papyrus was growing and hence the age of a piece of text estimated.[122]

Papyrus is strongly associated with Egyptian culture, although all the ancient civilizations around the Mediterranean used it as a writing surface. The papyrus sedge is a tall North African grass-like plant.

Rather than being brought into cultivation, papyrus was harvested from its native swamplands, including those on the banks of the Nile. Manufacturing papyri from papyrus sedge is a complex, messy process. Pith from the sedge's triangular stem is cut into long, thick strips that are laid side by side. These are then covered with a second layer of strips laid at right angles to the first, then soaked in water and hammered together. The hammered sheet is crushed to extract water, dried and then polished to produce a high-quality writing surface. Individual sheets would be glued together to make scrolls, or the fragile sheets carefully folded and bound into codices. In Roman times there was a complex nomenclature for different qualities of papyrus, ranging from the premier quality named after the Emperor Augustus to the lowest quality named after merchants.[123]

In moist climates the cellulose-rich sheets would readily decay, attacked by insects and mould, but in dry climates, such as the Mediterranean and Egypt, papyrus is a stable, rot-resistant writing surface; fragments thousands of years old have been recovered from Egyptian tombs. In 79 CE nearly 2,000 papyrus scrolls in the library of Julius Caesar's father-in-law were preserved at Herculaneum by ash from Mount Vesuvius.[124] However, the most famous discoveries of papyri have come from rubbish dumps of the ancient town of Oxyrhynchus, some 160 km south-west of Cairo, in the desert to the west of the Nile. Oxyrhynchus was a regional administrative capital and for a thousand years generated vast amounts of administrative bumf, including accounts, tax returns and correspondence, which was periodically discarded to make room for more. Over time, these dumps were gradually covered in sand and forgotten. But the papyri were preserved, creating a time capsule allowing astonishing glimpses into the lives of the town's inhabitants over the course of a millennium.[125]

Collections of documents that record information and ideas have frequently been viewed as subversive. For millennia, governments, despots and conquerors have used library- and book-burning as means

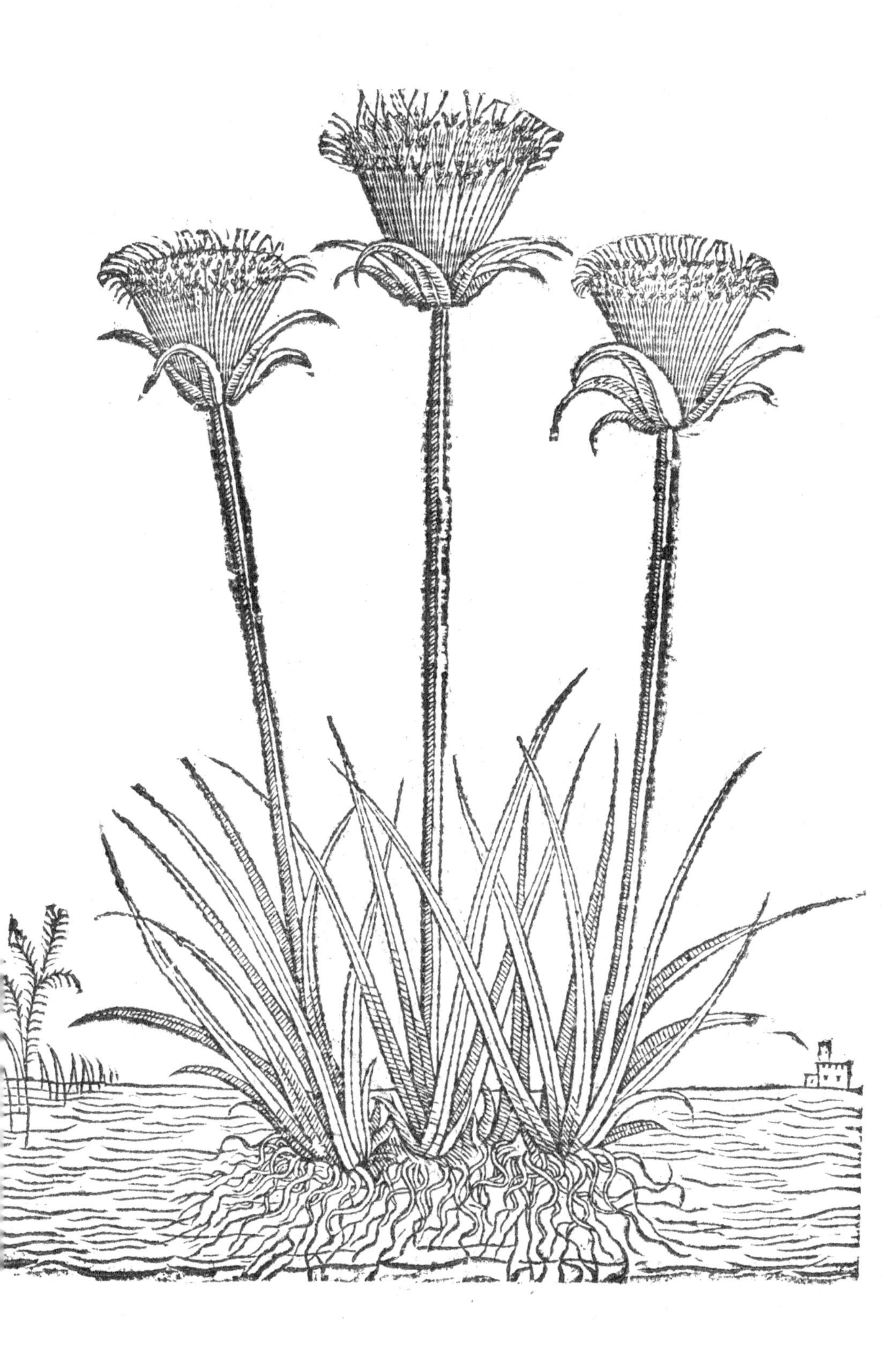

of destroying inconvenient evidence or trying to kill cultures and ideas that they found politically, morally or religiously unacceptable. One such calamity, the disputed destruction of the Great Library of Alexandria, and the papyrus scrolls and codices it contained, has been mythologized and come to symbolize the global loss of cultural knowledge.[126]

Besides their use in record-keeping, papyrus stems were used in many other aspects of Mediterranean life, such as rope-, sail- and basket-making, boat construction and as a food source. In 1969 the adventurer Thor Heyerdahl attempted to cross the Atlantic from Morocco in the boat *Ra*, to show that it was possible for mariners of the ancient Old World to cross the Atlantic Ocean. *Ra* was made from bundles of papyrus stems and modelled on an Ancient Egyptian craft.[127] As a marshland plant, papyrus sedge is a terraformer: it stabilizes soils and reduces erosion, while some investigations show its potential for water purification and sewage treatment.[128]

The papyrus sedge also gives us some of the language associated with knowledge and ideas. In Greek, papyrus pith is *biblos*, which gives us words such as 'bibliography' and 'Bible'. The word 'papyrus' is also Greek, being derived from *papuros*, the edible parts of the papyrus plant, and ultimately gives us the word 'paper'.

True paper was probably invented in China in the first century CE. Like papyrus, it was constructed from a meshwork of plant fibres, but the Chinese used fibres from the white mulberry tree, which yielded a tough, flexible material that could be folded, stretched, compressed and bent without being damaged.[129] (The white mulberry earns its own place in this book for another major contribution to civilization: see page 142.) The adoption of durable paper, and before that parchment and vellum, by Western cultures soon rendered papyrus obsolete. Over the last 2,000 years, laid papers made from the fibres of numerous plant species, including papyrus, have given way to pressed papers made from wood pulp.

Despite dreams of paper-free societies, Western cultures still use enormous quantities of paper, often in ways that it would be inconceivable to use sheets of papyrus. Sporadic attempts to revive the Western papyrus market since the mid-eighteenth century have failed. As a paper substitute, the role of the papyrus sedge in Western cultures has been superseded; papyrus is little more than a niche product for the tourist market. Today, the significance of papyrus for Western societies is in its history as the surface upon which our ancient ancestors recorded their lives, their art and their science; in the words of Pliny the Elder, it is the 'material on which the immortality of human beings depends'.[130]

Yew

Taxus baccata L.; Taxaceae

FIGHTING WARS AND CANCER

Among the accoutrements of Ötzi the Iceman, the 5,300-year-old mummified corpse discovered in the Tyrol, was a yew-hafted copper axe and an unfinished yew longbow stave.[131] Western civilizations' mystical, mythological, military and medical associations with yew, one of the most long-lived trees with which we share our planet, have been protracted. A yew spear, apparently more than 200,000 years old, was recovered from Clacton, on the East Anglian coast of England,[132] and among the animals depicted on the walls of the 17,500-year-old Palaeolithic cave complex at Lascaux, south-west France, are some brush strokes that have been interpreted as sprigs of yew. Yet yew's commercial and strategic importance has been so great that cultures from classical times to the present day have brought it to the brink of extinction across its entire natural range.

Yew is native but not endemic to Britain; it has a wide distribution from Europe, through India and China, to Japan.[133] The ease with which yew can be cultivated, and its range of forms, which reflects the species' immense natural variability, have made it a popular garden plant; today, more than 100 different yew cultivars are grown. From the mid-sixteenth century, French gardeners favoured yew to create highly geometric topiary shapes, taking the art of tree clipping to new extremes and so demonstrating their power over nature. The mid-seventeenth-century Keeper of Oxford Botanic Garden rivalled such absurdities when he clipped two yews at the garden entrance to be 'Gigantick bulkey fellows, one holding a Bill th' other a Club on his shoulder'.[134]

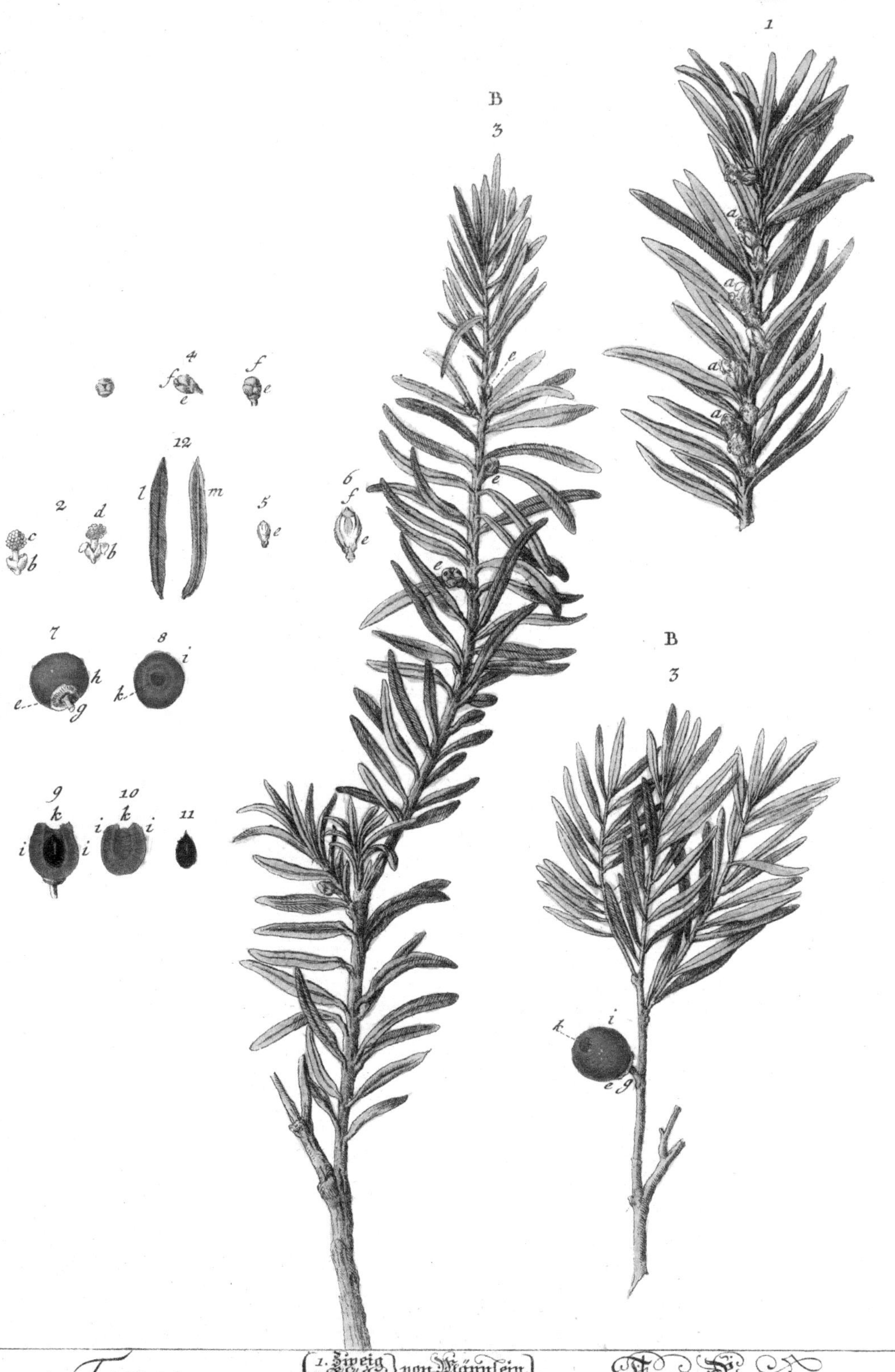

Taxus.
A. *mas.*
B. *femino.*

1. Zweig, 2. Blüthe — von Männlein
3. Zweig, 4–6. Blüthe, 7–10. Frucht, 11. Saame — von Weiblein
12. Blätter

Tax-Baum.
A. das Männlein.
B. das Weiblein.

Despite yew's familiarity, it often only attracts attention at specific times of the year. During the winter, its sombre green bulk contrasts with the bare branches of broadleaved trees. In early spring the branches of male trees are studded with golden male cones and the ground covered with drifts of yellow pollen. Then, in late summer, female trees reveal themselves, their canopies pricked with juicy crimson 'berries', which are in fact single-seeded, highly modified cones.

Yews can live to become ancient by the measure of our lives. In Britain, legend has it that the yew close to the church at Fortingall, Perthshire, sheltered Pontius Pilate and is nearly 5,000 years old; other estimates place the tree's age closer to 2,000 years old. Trying to age yews precisely has proven very difficult due to the complex manner in which they grow, but even with the lower age estimate the Fortingall yew is among Britain's oldest trees.

Given this capacity for longevity, it is unsurprising that a rich and elaborate lore, often associated with the language and customs of death and the afterlife, has developed around the yew.[135] Ancient Greeks considered yew sacred to Hecate, goddess of the underworld, necromancy and witchcraft, while in Britain one of the most frequent places to find yew is in churchyards. Some researchers have equated Yggdrasil, the Norse world tree, its roots in the underworld and crown in the heavens, with yew. People give names to important things, and many place names, in many languages, have their roots in yew. Eboracum, the Roman name for York, is derived from yew, while the Iberian Peninsula and the peninsula's longest river (Tejo) are both named after yew.

The English longbow, developed during the fourteenth century, transformed England's prominence as a Western European power. English and Welsh archers proved decisive at iconic battles such as Crécy, Poitiers and Agincourt during the Anglo-French Hundred Years War (1337–1453).[136] Yew is an excellent material for bow construction as, if correctly cut from a trunk, it provides a natural composite material: the elastic sapwood (on the outside of the trunk) performing well under tension and the heartwood (on the inside of trunk) performing well under compression. During the Hundred Years War, English and Welsh archers could rapidly send volleys of arrows more than 200 metres into French lines. Archers armed with yew bows were mobile, adaptable and deadly killing machines. However, training, equipping and maintaining archers came at vast cost, both financially and environmentally. The financial burdens fell on taxpayers and the communities that were required to supply trained archers. The environmental costs fell elsewhere.

England was not self-sufficient in yew, so an international trade developed, and for nearly 400 years the forests of Europe and beyond were stripped of yew. Gradually, as yew populations contracted, the quality and quantity of longbow staves imported into England declined. European leaders tried to encourage the creation of yew plantations, introduce extraction quotas and identify protected areas, but all was in vain. The environmental burden had fallen across the whole of Europe, despite the myth that the *English* longbow was made of *English* yew to defend the *English* way of life.

In England, the end came in 1595 when Elizabeth I proclaimed that her army would use shotguns rather than longbows. Some scholars have even suggested that, necessity being the mother of invention, the decline of yew spurred the development of the gun as a weapon of war.[137] The slow decline of the longbow, instrument of war, was also mirrored in the decline of a gentler instrument constructed of yew: the lute.

One of the facts most people know about yew is that it is poisonous. All parts of yew (except the fleshy crimson aril surrounding the seed) are indeed highly poisonous to mammals, producing a toxic cardiac cocktail, of which the most important compound is taxine B. When faced with defeat by Julius Caesar's army during the Gallic Wars, Cativolcus, co-chieftain of the north-eastern Gaulish Celts, the Eburones, took his own life with a decoction of yew, the tribe's totemic tree.[138]

However, plants toxic at one dose may have medicinal potential at lower doses. Consequently, yew extracts have been used for thousands of years against numerous complaints.[139] Yet these have not remained only part of traditional healing systems; they have become orthodox medicine's bestselling anti-cancer drug.[140]

Following an extensive search for compounds active against tumour cells, one potential candidate, paclitaxel, was isolated from the bark of a North American relative of yew in 1966. Paclitaxel was found to have novel actions on mammalian cells, and in 1992 finally got official clearance to be used in cancer chemotherapy. The hope, and hype, created by paclitaxel meant that yew had once again become a strategic resource, with enormous demands for yew bark. In 1966 global demand was about 106 kg; by 1990 this had risen to 39,000 kg; and in 1992 had leapt to 725,000 kg. The bark of one yew produces about 1 g of paclitaxel, and global yew populations could not sustain such assaults. But paclitaxel has such a complex structure that it is uneconomic to synthesize it artificially. Eventually a compound was discovered in yew foliage that could be converted, easily and economically, to paclitaxel. Yew clippings from sources worldwide, including gardens whose yews were originally planted purely as ornamental hedging and topiary shapes, came to contribute to the ever-increasing demand for paclitaxel.

Yews are disappearing from our landscapes through acts of gods and secular actions mediated by chainsaws and indifference. Whether Western cultures' yew heritage will extend into the future beyond place names and single, often isolated, trees that predate Christianity remains to be seen.

Rose

Rosa species; Rosaceae

SYMBOL OF PEACE AND POWER

An important, yet often overlooked, cultural role for plants is that they need have no other function than to look fabulous. Independent of their symbolic, medicinal, perfume, industrial and even minor food uses, roses have held their position as ornamentals in multiple cultures for thousands of years. Roses are among the most widely cultivated and loved garden plants. They bring pleasure to millions and employment to thousands, making hundreds of people rich in the process. Indeed, there are few places where Europeans settled where they have not tried to grow roses, however inappropriately. Breeders tinker with rose genetics to the same ends as their agricultural counterparts – to introduce disease resistance, for example, or to produce previously unknown character combinations. Would the history of genetically modified organisms during the early twenty-first century have been different if the first genetically modified plant on the market had been a true blue rose?

The eighteenth-century Swedish botanist Carolus Linnaeus took the classical Latin name for rose, *Rosa*, when he formally named the genus. He recognized twelve species, adding that 'species roses are hard to distinguish, more difficult to determine'.[141] Some seventy years later John Lindley in England was commenting on the amount of botanical attention roses were receiving; by the end of the nineteenth century the French botanist Jean-Michel Gandoger recognized more than 4,000 species from Europe and Western Asia.[142] The Belgian autodidact botanist François Crépin tried to make sense of this variation by arguing that hybridization among species was responsible.[143] He failed because he did not know enough about rose genetics; this came in the early decades of the twentieth century.[144] The readiness with which wild roses hybridize, the intricacies of their genetics and their popularity with gardeners mean roses show complex patterns of variation. The consensus today is the genus comprises some 190 usually thorny, shrubby species, distributed throughout the temperate and subtropical regions of the northern hemisphere: in the New World as far south as northern Mexico, in Africa down to Ethiopia and in Asia up to the Himalayas and across to the Philippines.[145]

In Europe, roses, particularly in the dog-rose group (*Rosa canina*), are common; evocatively scented, red, pink or white dog-rose flowers are a transient summer feature of hedges and waste places throughout Europe, but also in North Africa and Western Asia. The petals soon fall but the sepals that surround them often persist as the flower matures into fruit. Technically, rose fruits are the hard bony structures packed among the short stiff hairs inside brightly coloured rose hips. Indeed, the five sepals of the dog-rose calyx are the answer to the classical Roman riddle,

> on a summer's day, in sultry weather
> five brethren were born together.
> Two had beards and two had none,
> and the other had but half a one.[146]

Rose hips, rendered into syrup, have been used as food supplements, while crushed rose hips make effective itching powder for childhood pranks. However, the quality of rose-hip syrup – but not of itching powder – is highly variable, as vitamin C content varies with the source of the hips, when they are harvested, and cooking temperature and duration.

The fact that roses have been grown for centuries and readily hybridize, together with enduring geopolitical difficulties in parts of East and Central Asia, make the search for rose origins difficult. About seven rose species are thought to have made major contributions to the modern rose,[147] with breeders incorporating desirable traits such as

a. Rosa versicolor, Paſſe d'Angleterre, bunte Roſe. b. Rosa Prænestina versicolor. c. Rosa rubra Milesia flore pleno, Dammaſt-Roſen. d. Rosa præcox ſpinosa flore albo.

H.

brilliant yellow petals, winter hardiness, perpetual flowering and sweet scent into the narrow genetic background of modern roses.[148]

Before 1800 the so-called old-fashioned roses, the White, Cabbage, Moss, Damask and French roses, filled European gardens, bowers, literature, music and symbolism. At the end of the eighteenth century and beginning of the nineteenth century the splash of annual colour from these roses was extended by the development of perpetual-flowering hybrids, such as the Noisette and Bourbon roses, through the introduction of the 'China studs' from the Far East. Hybrids derived from perpetual-flowering and summer-flowering roses gave rise to the summer-flowering hybrid Chinas; from these hybrid perpetuals were in turn derived. In 1867 came the emergence of the first of the shrub roses, with their large, showy flowers, which would come to dominate the ornamental and cut-flower rose trade during the following century.

Rose breeders and hence gardeners have available to them a remarkable amount of variation: shrubs or climbers; scented or scentless; large- or small-sized; blowsy or discreet flowers. Each new cultivar, given a name for commercial, honourable, commemorative or sycophantic reasons, is not immune to extinction through the vicissitudes of the market or evolution. From the middle to the end of the nineteenth century, the number of commercially available old-fashioned roses declined steeply from about 800 to fewer than 90, with the numbers of available hybrid roses increasing during the same period.[149] However, Linnaeus was sceptical of the flower breeder: 'these men cultivate a science peculiar to themselves, the mysteries of which are known only to the adepts; nor can such knowledge be worth the attention of the botanist; wherefore let no sound botanist ever enter into their societies.'[150]

Artists have found inspiration in the rose. Some, such as Leonardo, Raphael and Botticelli, have used them as adornments to larger works, because of their Christian symbolism, or associations with love and romance. The seventeenth-century Dutch masters, such as Rachel Ruysch and Gerard van Spaendonck, produced elaborate still-life portraits of vases of the fashionable flowers of the day. Yet others, such as the Belgian flower painter Pierre-Joseph Redouté, who worked during and after the French Revolution, depicted roses as objects of botanical beauty in the gardens of both Marie Antoinette and the Empress Joséphine.

Political associations of the rose extend beyond paintings on Roman ceilings, and the confidentiality implied by the phrase *sub rosa*. In the latter half of fifteenth-century England, bloody dynastic squabbles periodically erupted between the royal houses of York and Lancaster over the English throne. Centuries later, with typical Victorian romanticism, these conflicts were christened the Wars of Roses after the White and Red Rose heraldic devices associated with York and Lancaster. The Lancastrian claimant (Henry Tudor, later Henry VII) defeated the Yorkist king, Richard III, at Bosworth near Leicester, and established the Tudor dynasty.[151] Henry VII united the Yorkshire and Lancastrian roses into the ten-petalled Tudor Rose; a nation's flower entangled with illusions of Englishness ever since. The discovery of Richard III's skeleton under a car park in Leicester in 2012 led to an undignified row over his remains, or perhaps their tourist value, between Leicester and York and a brief resurgence of partisan skirmishing between the red rose and the white. Modern political rose associations include the red rose as symbolic of international socialism, the White Rose anti-Nazi movement of 1940s Germany and the 2003 Georgian Rose Revolution.

In Ancient Greece, Theophrastus recommended chemical pruning – fire – to keep rose bushes under control and to encourage them.[152] Today, gardeners use other methods: the blade, pesticide, fungicide and fertilizer. However it is done, for the rose bloom to flourish above the prickle each year the gardener must be ruthless and wield absolute control – a metaphor not lost on politicians. Invitations into the Rose Garden of Downing Street or the White House imply more than just an opportunity to admire the blooms.

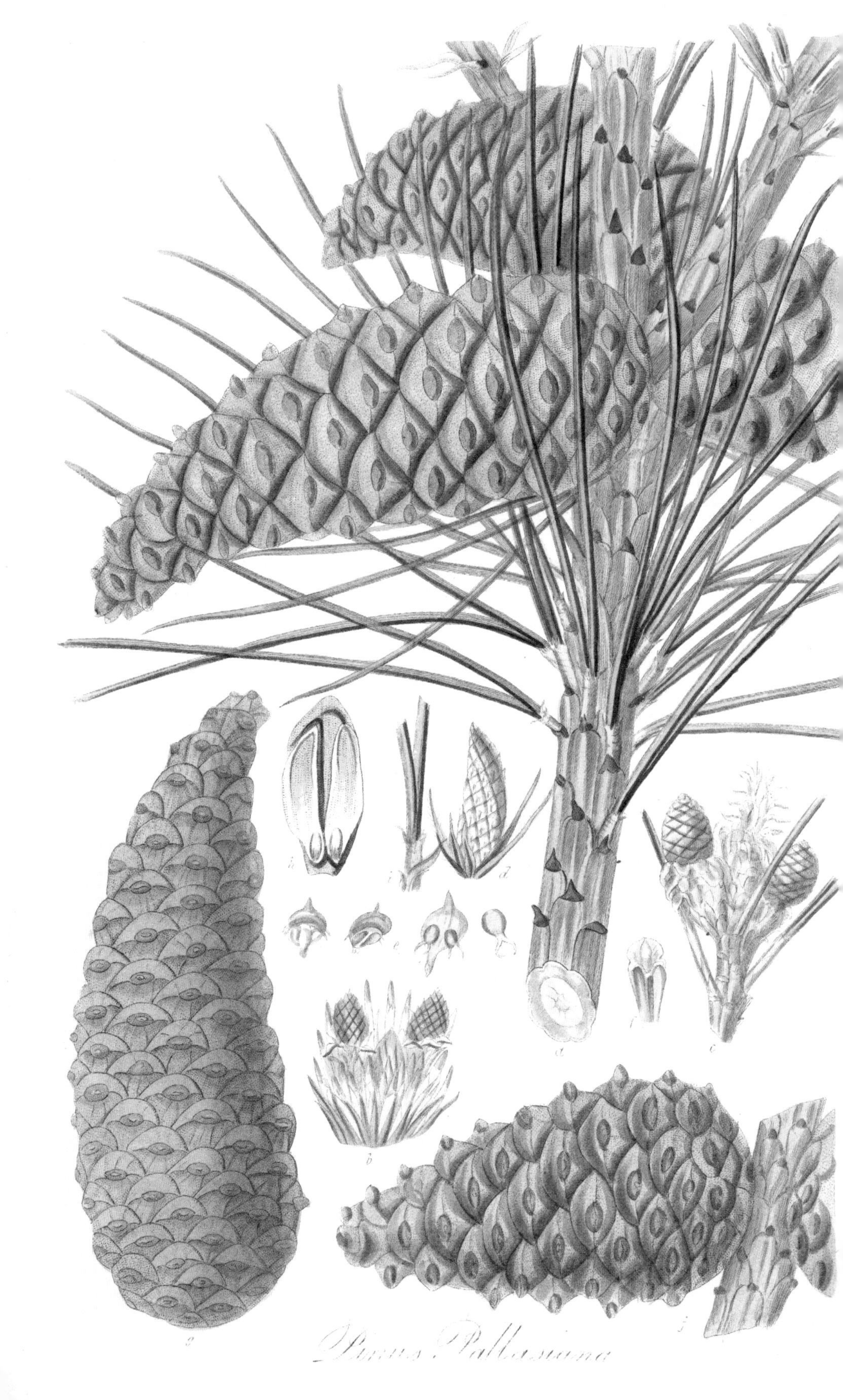
Pinus Pallasiana

Pines

Pinus species; Pinaceae

TIME CAPSULES

Trees respond to the stars – not in the way seventeenth-century astrological botanists would have us believe, but in how their annual growth rings are laid down. At the dawn of the twentieth century the American astronomer Andrew Douglass became interested in cycles of sunspot activity. He reasoned that if changes in solar activity affect global weather patterns these fluctuations would be reflected in plant growth responses, especially the patterns of the annual growth rings produced in the trunks of temperate trees. And so the science of dendrochronology, tree-ring dating, was born.[153]

Rings form in wood because of growth changes during the year; the inner part of a ring forms when growth is comparatively rapid (early wood), while the outer part of the ring forms when growth is comparatively slow (late wood). Furthermore, trees produce wide growth rings in favourable growth years but narrow growth rings in poor growth years. Consequently, growth rings give information both about the age of individual trees and about the growing conditions they experienced in particular years.[154] Pines have proved particularly valuable in reading tree rings, because of their longevity. Among their number are the longest-lived organisms on Earth. When the New World was discovered by Europeans, individual Californian bristlecone pines (*Pinus longaeva*) had already been producing seed for thousands of years; today, after more than 4,500 years, individual trees continue to flourish.

By piecing together tree-ring signatures in a temporal jigsaw of wood from many sources – the trunks of living trees, the stumps of dead trees, timbers inside old buildings and from archaeological sites, and from logs preserved in lakes, rivers and bogs – it is possible to produce highly informative sequences showing how trees have responded to climatic variation. The wide distribution of many common pines, together with their long history of human use, makes them ideal for producing long chronologies. Pines, in combination with other common trees, have enabled botanists to construct a tree-ring chronology covering the whole of the Holocene – that is, from nearly 12,000 years ago to the present day – for central Europe.[155] Besides helping us to understand how climates have changed, such sequences allow us to date artefacts at archaeological sites.

One of the defining features of pines is their cones, and the pollen within them also contributes vital information to the ongoing work of building up a picture of the planet in earlier epochs. Pine pollen grains have two air sacs to aid buoyancy, and the clouds of rot-resistant pollen released by pine trees can be carried great distances by the wind. Over time, under appropriate conditions, layers of pollen grains may build up in lake sediments. Because of the rot-resistant chemical (sporopollenin) they contain and distinctive sculpturing on their walls, pollen grains can often be identified even after being buried for thousands of years. A core of lake sediment is an ecological sample through time, where the pollen grains in each sediment layer reflect the prevalence of different plant species, allowing a picture to be drawn of how plant communities changed through time.[156] For example, pine data from pollen cores have shown how the distribution of pine forests has varied across the British Isles since the end of the last glaciation.

Another reason that pines are so invaluable to climatologists is their breadth of distribution around the world. As well as dominating the taiga, the band of forest around the globe between the Arctic tundra and more temperate southern latitudes, they are found in a wide range of habitats across the northern hemisphere, especially at high altitudes in tropical and subtropical regions.[157]

Using pine trunks and pollen grains as time machines to date man-made objects and understand millennia of climatic patterns is highly abstract. In practical terms, pines have proved to have immense commercial value. We have made use of pines for millennia as sources of timber, charcoal, resin and, in some cases, food and medicine. In their trunks are extensive networks of ducts that secrete volatile, aromatic, terpene-rich resins. Fossilized pine resin, under the name of amber, especially from around the Baltic, has been used in jewellery for thousands of years.[158] More prosaically, maritime nations needed wood to build ships and then naval stores (pitch, tar, turpentine and rosin) to maintain them at sea. Pitch and tar extracted from pine resin provided waterproofing and caulking for hulls and decking, and turpentine helped preserve rigging. Consequently, pines were a major strategic concern.

In the absence of sufficient home-grown resources, England relied on the Baltic states to supply naval stores, a situation that prevailed from the thirteenth century for around 400 years. But then political problems between England and Sweden, the War of the Spanish Succession (1705) and England's desire to exploit her colonies more effectively created a crisis in naval stores supply. With the payment of trade bounties, the English government convinced American colonials to exploit their pine forests for supplying naval stores. By 1725 the southern states of the Thirteen Colonies had taken over the Baltic's role as chief supplier of naval stores to the British shipbuilders; a situation that persisted until the American Revolution.[159] Following the revolutionary interruption, trade soon resumed and continued into the twentieth century.[160]

Until the end of the American Civil War, the arduous task of tapping trees, especially pitch pine, in the extensive forests of the southern states was the work of slaves – it was a bitter irony that the ships enforcing British abolition of the slave trade in the early nineteenth century were being kept afloat on the back of slave labour. When trees were too exhausted, old or uneconomic to tap, the industry simply moved on to new tracts of forest. However, despite the extent of naval store exploitation, the forest destruction caused was minimal compared to the extent of pine forests that disappeared under the plough, were burned on domestic hearths or felled in the construction of the railways.[161]

When vessels of iron and steel replaced wooden ships, the naval stores industry might have been expected to disappear. Instead, numerous new uses for pine products were discovered and have been supplied from forests in China and South East Asia. Turpentine is now a major solvent and raw material for the synthesis of complex industrial chemicals. Rosin, the hard residue that remains when fresh resin is heated, has become an ingredient in printing inks, varnishes, glues, paper coatings and soldering flux. It is also used in glazes for pills and chewing gum. Musicians and dancers will be familiar with rosin, too, as they use it to give a slight tackiness to bows of stringed instruments and to dancing shoes; rock climbers also find it useful to improve their grip.

Today, native pine forests continue to be exploited, but vast areas of land have been planted, often with non-native species, for both timber and paper pulp production. The use of exotic pines has created problems, with some species becoming invasive and threatening native plant communities. Such concerns, together with the impacts of pine plantations on soils, species diversity and landscape appearance, have led many organizations concerned with their 'green credentials' to join certification schemes that promote good forestry practice and stewardship.

While pine plantations represent pragmatic commerce, individual trees can acquire great symbolic value, particularly if gnarled and ancient. Their longevity gives them associations with wisdom, and in China the pine is traditionally recognized as one of the 'three winter friends', as it keeps its green needles through the winter. It is not hard to see why this long-lived, resilient tree has been adopted in many cultures as a symbol of endurance and immortality.

Reeds

Arundo and *Phragmites* species; Poaceae

TERRAFORMING TO MUSIC-MAKING

Landscapes, created by a combination of climate, geology and biology, especially plants, temper people's cultures. The inhabitants of the Mesopotamian marshes, wetlands in southern Iraq the size of Wales, are descendants of ancient marsh-dwellers who were part of the Sumerian cradle of Western culture. Their lives, like generations before them, have been shaped by this watery landscape, although the West remained largely ignorant of it until British explorers Wilfred Thesiger and Gavin Maxwell wrote about their expeditions to the region just after the Second World War.[162]

The marshes are dominated by standing water and the common reed (*Phragmites australis*). This is a grass with long, broad, spear-shaped, blue-green leaves held on robust stems that can reach 3.5 metres in height.[163] The stems arise from branching rhizomes that creep through the mud at the bottoms of rivers, lakes, ponds, marshes, swamps and fens. Common reed produces dense growth and spreads rapidly across large areas; even small fragments of rhizome can quickly establish a colony. What is more, common reed will grow in fresh or brackish water, on acid or alkaline soils. As might be expected from such a highly adaptable species, common reed has a wide distribution; it has even been claimed as the most widely distributed flowering plant on the planet.[164] Of course, these very characteristics mean that in appropriate habitats it can become a serious weed, with single genetic types covering large areas.

Since ancient times, the term 'reed' has been applied to many different plant species, most of which appear like tall grasses and are found near water. One of the most imposing species is the giant reed (*Arundo donax*), a grass that has been used for millennia in Western Asia and the Mediterranean, whose stems can reach 10 metres tall and 3 centimetres in diameter.

Common and giant reeds are multi-purpose species and have been used for thousands of years as basic construction materials for building houses, thatching roofs and making baskets and musical instruments.

Despite the multiple day-to-day uses of reeds, reed-covered land is sometimes viewed as malaria-ridden and of limited agricultural value. Consequently, schemes have been encouraged to drain these areas and convert them to 'productive' lands to feed ever-expanding populations or to satisfy the economic and political aspirations of landowners. During the seventeenth and eighteenth centuries, reed-dominated lowlands in the Netherlands and England were drained to provide agricultural land.[165] In the case of the Mesopotamian marshes, draining began in the 1950s to reclaim land for agriculture and oil exploration.[166] However, by the early 1990s civil engineering projects had been subverted by Saddam Hussein's Iraqi regime, to become collective punishment for perceived rebellions by the Marsh Arab population.

The positive economic and social benefits of landscape transformation must be balanced by the risk of exposing fragile soils to erosion. Fertile soils, slowly created over millennia, are essential if land is to be used productively. However, the combination of water and wind can rapidly erode soils. About one-third of the planet's land area has been converted to agriculture, although in the past forty years approximately one-third of this agricultural land has had to be abandoned because of soil erosion. At first sight, soil erosion may not appear a cultural issue, yet it has dramatic effects on communities at all levels, through economic collapse and social upheaval.[167] The fate of empires and the ability to feed their populations across the Old and New Worlds have been associated with fertile, productive soils; loss of these soils cannot be mitigated readily by technological fixes. Reeds typically develop

Arundo Phragmites. Common Reed. ♃

C. Mathews, Del. & Sc.   Pub.d by W. Baxter Botanic Garden Oxford. 1840.

rapidly and have roots and shoots that bind soil particles and limit soil movement. This makes them important wetland engineers and terraformers, stabilizing the banks of watercourses, modifying patterns of water flow and even acting as flood defences. Reeds are also used as biological scrubbers to aerate and clean polluted water, enabling, for example, the treatment of waste water and industrial effluent.[168]

Reed biology means reed-dominated communities can be recreated in flooded areas. Until the 1960s the English Broads were thought to be a natural feature of East Anglia. However, we now know the Broads are artificial, produced by flooding medieval peat excavations.[169] In the first decades of the twenty-first century there was hope that large parts of the Mesopotamian marshes, and their displaced people, would return, but recently this has been tempered by the twin challenges of water scarcity in the region and the projected effects of climate change.[170]

The structure of reed stems includes a high proportion of cellulose, and industrial uses for cellulose have been proposed since the eighteenth century, for example for the production of low-quality paper. However, it was in Italy in the 1930s and 1940s that giant cane cellulose became of national importance as the raw material for the production of the synthetic fibres rayon and viscose.

Interest in reed cellulose has returned in the twenty-first century with concerns about carbon emissions and the search for biofuels. Today, most fuels come directly or indirectly from fossil fuel supplies. If coal, oil and gas can be described as stemming from historical photosynthesis, biofuels are derived from present-day photosynthesis.[171] Cellulose-rich reeds have been proposed as biomass crops for energy production, although the high silica content of grasses means there can be problems with clinker build-up if they are burned. Consequently, chemicals and enzymes are being explored that will convert cellulose in reed cell walls into liquid fuels efficiently.[172] However, the question of whether biofuels are ecologically sustainable, together with the economics and ethics of biofuel production, remain topics of great debate and complex analyses.

One of the most famous and, to most westerners, important uses of the giant reed stem is to make musical instruments. The stem is a hollow cylinder divided at intervals by horizontal walls. Cutting the stem into different lengths creates a simple musical instrument: panpipes. More significantly, fine stems may be fashioned into reeds for woodwind instruments, such as the clarinet, bassoon and oboe. A reed stem is constructed of three concentric tissue rings: a hard, waxy epidermis, a band of thick-walled fibres and a thick inner layer of vascular bundles. The structure of these layers is essential in determining the quality of musical reeds. Some of the world's most sublime music has been given flight by musicians blowing across a piece of grass stem.

a. Quercus, Chêne, Eichen-Baum.
b. Quercus calyce echinato glande majore seu cerrus, Chêne amaire,
Bittere Cer-Eiche.

Oak

Quercus species; Fagaceae

THE TREE THAT LAUNCHED A THOUSAND SHIPS

In Western cultures, oak is the botanical equivalent of the pig: both have ancient, complex, pervasive, multi-purpose associations with the people they serve.[173] However, whereas humans have transformed wild pigs through domestication, oaks, like many trees we use, are, at best, only semi-domesticated. Whatever the truth of the old saw that an oak takes 200 years to mature, stands for 200 years and dies for 200 years, by our standards oaks live a very long time. The length of their lives imbues them with mythological, symbolic and mystical charm that is reflected in the stories and legends surrounding them.

The genus *Quercus* comprises some 600 species spread across the world's northern temperate regions. Most are forest trees, with distinctive acorn-like fruits held in cups. Western peoples have usually exploited the native oaks they find around them, although particular species may have been imported for special purposes. Oak, *carvalho*, *roble*, *chêne*, *Eiche*: these are words associated with quality and craftsmanship, attracting the acclamation of millions across Western cultures – in the modern argot, they have high brand recognition. However, oak is a slippery common name. By tradition, true oaks belong to the genus *Quercus*, and perhaps its close relatives. Yet 'oak' is often applied to other unrelated genera and species: Australian oak (*Eucalyptus regnans*), poison oak (*Rhus radicans*), she-oak (*Casuarina* spp.), silky oak (*Grevillea robusta*) and Spanish oak (*Inga laurina*), to name but a few. One explanation for such a proliferation of oak-based names might be that European immigrants saw resemblances between the trees of their homelands and their adopted lands, perhaps because of leaf shape, ubiquity or properties. Less charitably, the cachet of the word carries high financial value for products bearing the name, which invites deceit.

About 16,000 years ago ice advanced across the northern hemisphere as the climate cooled. Palaeobotanical and genetic evidence shows oak retreated south, together with many other plants and animals.[174] In Europe these northern refugees eventually found new homes in isolated pockets (refugia) of the northern and eastern Mediterranean. During the confinement of millennia, mutations in oak DNA accumulated and populations in different refugia gradually diverged from each other, although not sufficiently to be considered different species. When the Ice Age ended, about 10,000 years ago, oak emerged from the refugia to recolonize the ice-scoured land. Sequencing oak DNA reveals the genus's recolonization history. Ironically, long before Europe was arbitrarily divided by political boundaries, oak recolonized Britain from the Iberian Peninsula, via northern France; so the oak the British used to build their ships and armoury to keep the Spanish and French at bay for centuries ultimately came from their own lands.[175]

Like other trees, oaks lay down each year a thin layer of new tissue, producing a growth ring in the wood (see PINES, page 95). Splitting an oak trunk opens up a crack through time that can reveal how the tree grew and the environment in which it grew. Such details are not only of great academic interest; they have great practical interest since growth conditions will determine oak properties.[176] Despite different oak species being used in different places having differing properties, common themes emerge as to the uses to which this most versatile of trees may be put. Perhaps the most obvious is as timber.

The importance of oak as a timber species is all around us in architecture, in formal and rustic furniture, and in household goods: the timber frames of Tudor buildings, the hammer-beam roofs of churches, and chairs, tables, benches, beds and barrels. More famous still are the ships which, from the sixteenth century until the age of steam arrived in the mid-nineteenth century, took Europeans across the globe. Ships transported goods and ideas, noble and ignoble, along with diseases, between cultures; the world began to shrink. Ships, particularly those of the British Navy, dominated the world's oceans and exercised 'hard power' in the pursuit of imperial ambition. 'Hearts of Oak are our ships' sang British sailors as they sailed off to vanquish the world, 'We'll fight and we'll conquer again and again.' Similar ships also exercised 'soft power', changing hearts and minds, through trade that bound distant countries and peoples together. Henry VIII's warship *Mary Rose* is estimated to have consumed 600 oak trees in its construction, while Horatio Nelson's flagship HMS *Victory* needed nearly 6,000 trees, the vast majority of which were oak.[177] These vessels, the elite of their day, were just part of fleets of hundreds of smaller ships, both military and commercial, constructed from oak. Not only was high-quality oak essential for ship construction, but there had to be a constant supply of oak suitable for their repair.[178]

Oak timber and branches are also converted to charcoal, creating fuel for metalworking and cooking, while smouldering oak chips are important in food preservation. Oak is also an important flavouring, whether as smoked food or as the flavour imparted to whisky as it is aged in oak barrels.

Oak bark, often a by-product of forestry operations, is tannin-rich and essential for the business of the leather tanners. Until barks of certain tropical trees became widely available, oak was the only bark suitable for tanners to use. By the mid-nineteenth century the demand for oak bark was so great that it became more valuable than the timber. The bark of some oaks, most notably the cork oak of the western Mediterranean (*Quercus suber*), produces the cork of commerce. For the tree, cork is both a waterproofing agent and a protection again fire damage. Humans have usurped both of these properties to stopper wine bottles and provide fireproofing, and intricate management systems have evolved for its sustainable production. In the cork oak forests of Spain and Portugal the trunks of selected trees gleam blood red at harvest time, where bark has been removed and living layers beneath exposed.[179]

Oak galls, the apple-like swellings that oak trees produce in response to gall wasps laying their eggs in the young branches, are also tannin-rich. Mixing ground oak gall with iron nails dissolved in vinegar produces an indelible brown-black ink. Iron gall ink was used for nearly 2,000 years, until the development of synthetic inks in the late nineteenth century. Records of Western civilizations are written in gall ink on papyrus, vellum, parchment, paper and even oak strips.[180]

Traditionally, pannage, the practice of releasing domestic pigs into woodlands to root for fruits such as acorns, has been an important part of the management of European oak woodlands.[181] Today, ham from pigs raised in oak woods carries a cachet and is priced accordingly. Acorns were once an important human food, especially in times of famine, around the Mediterranean, and were pressed into service during the Second World War, ground into ersatz coffee.[182] Today, they are rarely used, since to make them palatable they usually require considerable processing to remove the bitter chemicals in the seeds.

Oak remains an iconic tree of Western cultures, and is often strongly associated with nationalistic sentiment. Today, oak forests, albeit highly fragmented by the requirements of modern human environmental exploitation, are areas where people extract livelihoods. However, they have also become areas for recreation and for the conservation of the numerous plant, insect and animal species associated with oaks. Isolated oaks, such as the Wyndham Oak in Dorset, have become symbolic of changing landscapes, often fortunate exiles from formerly larger forested areas.

Apple

Malus domestica Borkh.; Rosaceae

OUT OF ASIA

The apple is a remarkable tree. It has become part of the diet and lives of most peoples who have had contact with Eurasians. Practically, the apple is a source of sugar- and vitamin-rich food that stores well during lean times, and apples are the raw materials for brewing cider and distilling calvados. Unsurprisingly, the apple's usefulness and ubiquity have produced a plethora of customs, myths and legends.

Apples were familiar across the classical world, and laden with the symbolism of sexuality and temptation, but also with chaos and destruction. Such symbolic baggage affected the fruit's reception, especially once it became entwined with the Genesis myth of the Fall of Man. The Romans were particularly effective at disseminating apples across their empire. With the fall of Rome, apple cultivation was taken up by monasteries, so that by the thirteenth century the number of apple cultivars grown across Europe had soared. Protestant migration from Europe disseminated apples to North America and created the folk hero Johnny Appleseed. Iberian adventurers populated their Central and South American empires with apples, while the British scattered them across their own colonies. Victorian apple connoisseurs produced lavishly illustrated, exorbitantly priced pomological volumes that celebrated the diversity and utility of apples. The apple, with its gastronomic, symbolic, religious, decorative and patriotic appeal, is culturally and biologically adaptable.

The domesticated apple is part of a small genus of approximately thirty species distributed across the temperate regions of the northern hemisphere. The precise number of species is difficult to determine because of high levels of variability and promiscuity among apples, and their widespread cultivation. Closely related species readily cross, producing complex hybrids, while cultivated and wild apples frequently cross to produce highly variable wildings.

People have speculated about the origins of the apple for centuries. Over the last century, through the combination of intensive fieldwork, experimental and commercial breeding, and molecular genetics, science has greatly enhanced our ability to disentangle the intricacies of the apple–human relationship and the origins of domesticated apples.

The apple we eat is not European; it is native to Central Asia. The Central Asian wild apple, *Malus sieversii*, is a common species of the remarkable Central Asian fruit forests.[183] These temperate forests once extended west from China in a narrow strip along the Pamirs and Tian Shan to the Caucasus and the Black Sea. However, numerous destructive policies, especially during the Soviet era, have confined this precious forest to a few fragments on the border of Kazakhstan and China, and parts of Kyrgyzstan; fragments which continue to disappear. The Central Asian wild apple is naturally very variable in tree height and form, and in fruit size, shape, flavour and colour; some wild trees even produce fruits of sizes and qualities that would be acceptable to Western supermarkets. Out of this pool of natural variation, people selected and propagated their favourite types, which some four to ten thousand years ago produced a distinct species; the Central Asian wild apple had been domesticated. Over the next four millennia, as the apple moved west, perhaps travelling with nomadic horsemen and people trading along the Silk Roads, further selections were made as the domesticated apple picked up genes from the Western Asian wild apple, *Malus orientalis*, and the European wild apple, *Malus sylvestris*.[184] Such unconscious efforts, together with the conscious efforts of generations of apple breeders since the early nineteenth century, have produced thousands of named apple cultivars.

a. Malus domestica, Pom̃ier, Apfelbaum. *b. Mala dulcia,* Süßling.
c. Mala angulosa, Zapfapfel. *d. Mala fragrantia curtipendula,*
Borsdorffer Apfel. *e. Mala rubentia,* Rübiner Apfel.

s

Trying to propagate particular apple cultivars from seed is a waste of time. A single apple tree will not mate with itself but produces seed by crossing with nearby individuals; hence, each seed from a single tree will have different parents. In order to produce identical trees, breeders and growers need to use a vegetative propagation technique, such as grafting. Grafting, where the scion of one tree is attached to the rootstock of another, appears to have been invented about 4,000 years ago in Mesopotamia. The Greeks established apple propagation techniques in the third century BCE, and the Romans later scattered the technology across their empire. Grafting, and the ability to bulk up particular apple types, probably played a key role in establishing the apple as a major fruit crop. In addition, their popularity as a commercial crop increased once a way was found to restrict the size of trees so that they could be easily maintained and harvested. Rootstocks were the key to solving this problem. By the mid-twentieth century a small number of popular dwarfing rootstocks had been selected and established as preferred rootstocks. Today, few trees, either in commercial cultivation or planted in gardens, grow on their own rootstocks.

Globally, apple production has risen steadily over the last fifty years, with China currently the world's largest producer. The desire of many consumers and retailers for blemish-free apples that store well and are easily transported means only a handful of apple cultivars are widely marketed today – for example, 'Braeburn', 'Fuji', 'Golden Delicious' and 'Granny Smith'. During the last century, as agriculture became more intensive, the low economic returns associated with apple production meant orchards were grubbed up across Europe. However, movements have emerged promoting the significance of local apple diversity and heritage.

In the absence of mutation, grafting means apple cultivars are passed through the centuries unchanged. As long as a horticulturalist takes the effort to propagate a particular type, we can see, smell and taste the apples that were familiar to Shakespeare and Newton. But apple cultivars cannot be stored in seed banks, only as collections of living trees. Living plant collections are insurance policies, although the cost of maintaining the premiums is high. The diversity of apple cultivars, carefully maintained by centuries of grafting, and the genes from Eurasian wild apples are vital resources if apple production is to be maintained in the face of persistent global threats, such as pests and diseases.[185] Cash-strapped public funders often try to balance intangible potential future benefits against immediate fiscal and political returns; plant collections are seen as expendable. However, if these apple collections are to remain important for apple breeding they must expand, especially by the addition of rootstock variants, local and ancient cultivars from beyond Western Europe, and the Eurasian wild apples, of which there are too few samples.

a. Piper nigrum, Poivre noir, schwartzer Pfeffer.
b. Piperis ramus, Pfefferzweig.

Pepper

Piper nigrum L.; Piperaceae

KING OF THE SPICE TRADE

Rare, highly desirable items command vast prices. Five hundred years ago the contents of the now commonplace household pepper pot would have been worth a small fortune, reflecting the esteem in which the pungent little peppercorns were held and the complexities and dangers involved in getting them from Asia to Europe. The pepper trade marks the earliest and most well-documented trade links between Europe and Asia, and exemplifies how spices caused fortunes to be made and lost, drove discoveries and made empires.

Pepper is the most widely traded spice on the planet. In 2022 the export value of approximately 8 million tonnes of pepper to producer countries was over US$2 billion. Vietnam, the world's largest pepper producer, earned more than US$850 million in exports, although at least one pepper commentator dismisses Vietnamese pepper as tasting like 'chewed tobacco' and having 'no warmth or depth of flavour'.[186]

Pepper, the hard, dried fruit of an evergreen Southern Asian vine, was known in antiquity; the philosopher Theophrastus, in fourth-century BCE Greece, mentions it, for example.[187] By the first century CE Romans were routinely crossing the Arabian Sea to India. Pliny the Elder bemoaned the drain on Roman coffers because of the millions of sestertii flowing annually from the spice (including pepper) trade into the hands of Indian traders.[188] Roman trading vessels, laden with Indian goods, would travel along the Arabian Peninsula to Aden and up the Red Sea to Egypt, then overland to Alexandria, the Mediterranean and Europe. This route dominated the pepper trade until the fall of Constantinople in 1453. However, the Indian traders were not only supplying pepper to Europe; Persian and Muslim middlemen were trading significant amounts of it east into China and supplying India's own internal market.[189]

Piper nigrum is a woody vine of tropical forest shade, and grows best in areas of high rainfall. Tiny flowers, clustered together into long catkins, eventually develop into small, bright red spherical fruits, tightly packed along the catkin. Until the gradual establishment of pepper plantations across the tropics from the late eighteenth century, *Piper nigrum* and pepper production were restricted to the Malabar Coast of India (present-day Kerala). Given the antiquity of human–pepper associations in India, it is hardly surprising differences should have evolved between cultivated and wild forms. Wild pepper tends to have separate male and female plants and reproduces sexually, so only a proportion of the vines will produce fruit and any offspring will have different characteristics to their parents. In contrast, cultivated pepper tends to have hermaphrodite flowers and is propagated asexually, meaning all plants are capable of producing fruit, and any cultivated types well adapted to particular areas can be readily propagated.[190]

Black, white and green peppercorns come not from different species, but from different processing. Black peppercorns are unripe fruits that have been briefly cooked then dried. To produce white pepper, ripe fruits are retted (that is, soaked to break down the hard casing) then cleaned to separate seeds. Green peppercorns are unripe fruits that have been dried or pickled. Processing differences produce flavour differences, and some people have strong preferences, but it is the combination of genes, growth conditions and processing that produces high-quality pepper, not solely the colour.

Yet not everything traded as pepper is *Piper nigrum*. Other *Piper* species are frequently found, particularly before the eighteenth century; for example, Indian long pepper (*Piper longum*) and Indonesian tailed pepper (*Piper cubeba*). A peppery flavour has also lent the name to quite unrelated plants, including Brazilian pink pepper (*Schinus molle*), tropical African Guinea pepper (*Xylopia aethiopica*) and West African Melegueta pepper (*Aframomum melegueta*).

Besides its role as a food additive, pepper has medicinal value, although whether pepper-based products are effective is a matter of active debate. The pungent alkaloid in pepper, piperine, is promoted as a natural insecticide.

Pepper, together with spices more generally, was fascinating to Europeans because it worked on the olfactory senses in ways few native European plants could. The rarity of spices, and the risks and costs associated with transporting these fragrant fragments from the other side of the world naturally lent them cachet.[191] Furthermore, they pandered to eternal beliefs that medicinal panaceas and sacred insight could be found in the exotic. However, the most important reason for fascination with pepper was perhaps a worldly one – spice traders accrued vast wealth.

The spice trade is well documented, attracting generations of social and economic historians because of its importance in establishing the wealth and power of Venice and driving Iberian exploration of sea routes to the Indies in the fifteenth century. As a result of the wealth it generated, pepper influenced art. Pepper consumption became a status symbol; people wanted to show off that they could afford it. Romans, who made liberal use of pepper in their cooking, made elaborate pepper pots, such as the Hoxne Hoard Pepper Pot discovered in Suffolk in Britain in 1992,[192] and the merchants of the Dutch Golden Age commissioned symbol-rich still-life portraits of the luxuries they could afford, including pepper.[193] The fabulous salt cellar that the sixteenth-century goldsmith Benvenuto Cellini created for François I of France is a table piece for both salt and pepper, with a golden temple to hold the precious condiment.

Governments taxed pepper, so the fiscal waltz between the excise and smuggler became ever more intricate as importers conceived ever more elaborate wheezes to evade their liabilities. High-value products also attract criminals to pass inferior material off as the real thing. Even in early-nineteenth-century Britain, when pepper cost a mere fraction of what it had been in previous centuries, counterfeiters were at work. In his compendium of food adulterants, the German chemist Frederick Accum described the efforts to which pepper counterfeiters would go to dupe their customers. Black peppercorns were fashioned from linseed residue mixed with clay and chilli pepper, or an 'injurious mixture of rape and mustard'.[194] Ground black pepper was sold mixed with pepper dust (sweepings from the pepper warehouse) or an 'inferior sort of this vile refuse'[195] called sweepings of pepper dust. The value of black peppercorns could be increased by transmuting them into white pepper by steeping them in seawater and urine. Early trading standards legislation attempted to protect producers and consumers, and, perhaps more importantly to them, government tax revenue, yet fines of up to £100 did not deter the fakers and fraudsters.

Once transportation costs fell, so did prices, and the trouble of counterfeiting was no longer worth the return, although low-quality peppers may still be passed off as high-quality because of the premium on specific regional peppers.

As pepper lost its value and its status, the meaning of terms associated with pepper began to change. Pepper had once been a currency and means of both storing and exchanging wealth. When the Visigoth Alaric I threatened Rome in 408 CE, part of the ransom paid to prevent Rome being sacked included 3,000 pounds of pepper.[196] (This was only a short-lived remission; Alaric's forces sacked Rome two years later, beginning Rome's decline as a military power.) By the mid-nineteenth century, peppercorns had become associated with the idea of a mere trifle; a peppercorn rent.

Carrot

Daucus carota L.; Apiaceae

DUG FOR VICTORY

The family of plants known botanically as the Apiaceae includes familiar foods and spices such as carrots, parsnips, celery, parsley, coriander, cumin and dill, but also such poisonous plants as hemlock, cowbane and water dropwort. Distinguishing between such plants may make the difference between life and death, so learning to identify plants accurately must have had a high priority in our evolutionary history. The processes by which we learnt to discern toxic from non-toxic and nutritious from non-nutritious appear to have been through observation of animals, trial-and-error experimentation and serendipity. Furthermore, being able to communicate about the specific properties of plants depends on giving them unique names. For peoples as diverse as the Egyptians and Assyrians through the Chinese and Indians to the Greeks, names were crucial. Indeed, one of Adam's first challenges in Eden, according to Genesis, was taxonomic: to name 'cattle, ... fowl of the air and every beast of the field'.[197]

Just as people learnt to take advantage of plants for their medicine, they appreciated their starch and sugar reserves as food stores. Carrots came to be valued, and cultivated, for the satisfying sweetness of their swollen roots. Carrot plants are biennials; they take two years to mature. In the first year the cells of the outer layers of the taproot swell with stored sugars. In the second year these sugars are used to fuel the formation of flowers and fruits. However, cultivated carrots rarely flower since they are harvested at the end of the first year when taproot sugar contents are at their highest. Inside specially constructed root stores or left to overwinter in the ground, carrots will maintain their sugar contents and nutritional value for several months.

Classical Greek and Roman writers were aware of the medicinal and culinary possibilities of carrots but did not clearly distinguish them from parsnips, a confusion maintained into the early modern period. For example, the Greek philosopher Athenaeus (second century CE) reported that carrots (or possibly parsnips) were 'very nourishing and fairly wholesome, with a tendency to loosening and windiness; not easy to digest, very diuretic, calculated to rouse sexual desire; hence by some it is called love-philtre'.[198] Today, there is little chance of confusing bright orange carrots with dirty-yellow parsnips, but, although some of the cultivated carrots illustrated by first-century physician Dioscorides in *De Materia Medica* had orange taproots, the cultivated carrots known to many classical and historical authors were purple or yellow.[199]

Despite this confusion over carrots and parsnips, classical authors made clear distinctions between the roots from cultivated (manured) plants and wild ones. The former made better food, while the latter made better medicine: 'the Wild are most effectual in Physick, as being more powerful in their operation than the Garden kind',[200] observed the seventeenth-century physician Nicholas Culpeper.

Daucus carota is a very variable, widespread species naturally distributed from the Atlantic coast of Britain and Ireland through Europe and the Mediterranean to Central Asia. The species is readily recognizable by its highly divided carrot-scented, fern-like leaves and clusters of tiny white flowers, arranged in parasols, arising from a nest of finely divided bracts. Underground, wild carrot has a small, tough, highly branched white taproot. In contrast, cultivated carrots have swollen, unbranched taproots in a rainbow of colours – purple, yellow, red, orange, black.

There are two classes of pigments in carrots: anthocyanins, which are purple, and carotenes (orange beta-carotene, yellow xanthophyll

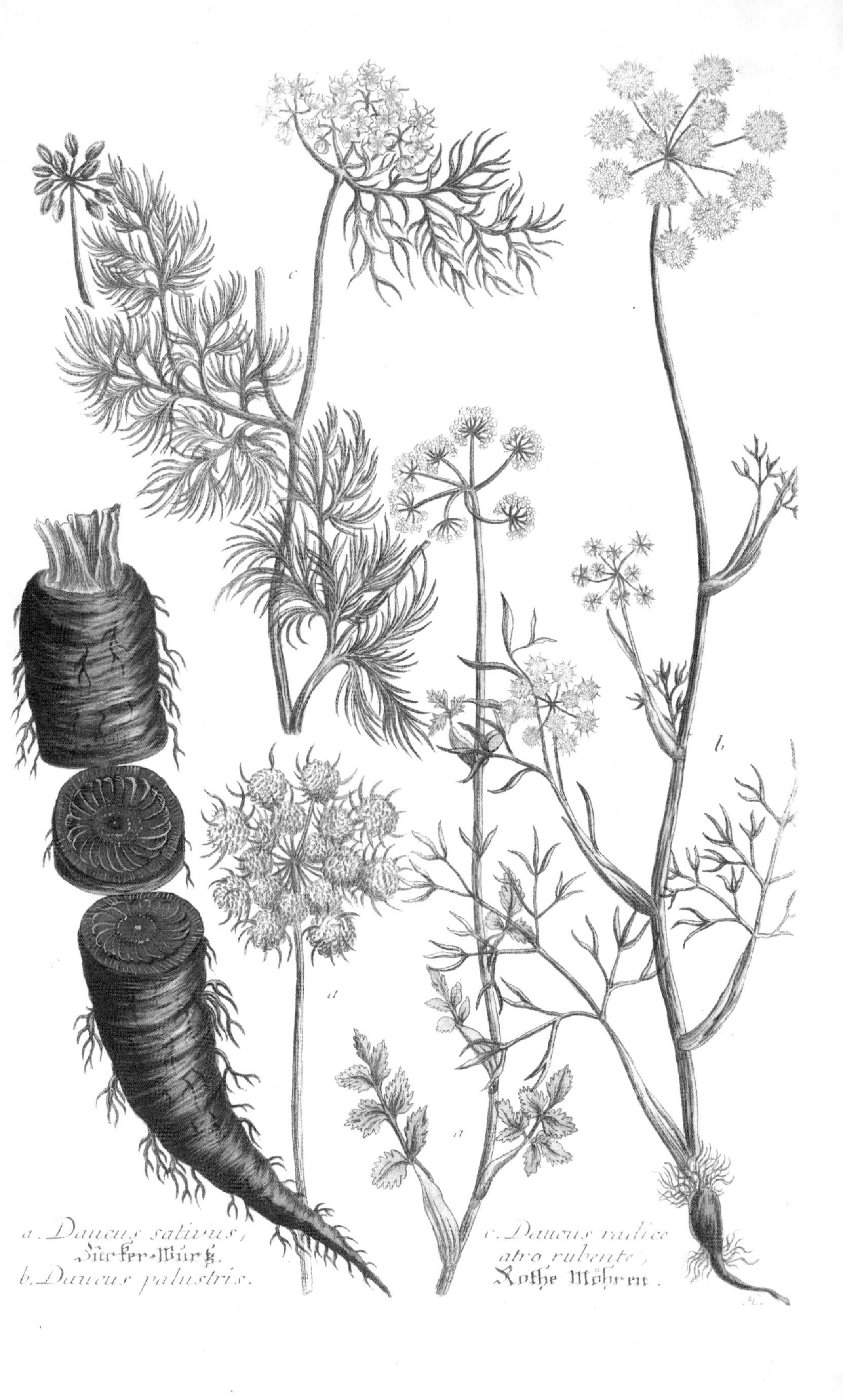
c
a
b
a
a. Daucus sativus,
Zucker-Würtz.
b. Daucus palustris.
c. Daucus radice
atro rubente,
Rothe Möhren.

and red lycopene). All have valuable health-giving properties. Anthocyanins are powerful antioxidants; carotenes are essential sources of vitamin A, an important component of pigments found in the retina of the eye. Without sufficient vitamin A we are likely to suffer visual problems such as night blindness.[201] Purple carrots are also rich in carotenes, though their colours are masked by the anthocyanins, but these are water-soluble; when boiled, purple carrots turn a rather unattractive brown. In contrast, xanthophyll-rich yellow carrots retain their colour when cooked (the carotenes are oil- not water-soluble) – making them an attractive alternative to purple carrots.[202] In addition to colour, other differences among the multitude of carrot types include the shape, size, sweetness and bitterness.

Carrot seeds have been recovered from prehistoric sites up to five millennia old, although whether they were used as medicines or spices, or merely contaminants, is unknown.[203] Historical and genetic data and the distribution of domestication features indicate domesticated carrots originated in Central Asia.[204] In the early twentieth century the Soviet geneticist Nicolai Vavilov was emphatic: in Central Asia the 'wild carrots … practically invited themselves to be cultivated'.[205]

By the tenth century a vast array of white, yellow, purple and black cultivated carrots were found in the region of Afghanistan.[206] Modern carrots were selected from wild carrots in the east of the species' range. Their ubiquity across Europe is due to the spread of these cultivated carrots, not, as might be expected, the result of independent domestication wherever wild carrots were found in Europe. During the tenth and eleventh centuries the cultivated carrot spread into Asia Minor, and during the twelfth century was taken to the Iberian Peninsula by the Moors. In the fourteenth century it spread through north-west Europe and finally entered Britain in the fifteenth century. As it continued its westward journey, eventually into the Americas, it also spread east into China and Japan.

During the seventeenth century the kaleidoscope of carrot colours in Western Europe was gradually reduced to one, and European carrots became stereotypically orange with the popularity of 'Long Orange' and 'Horn' varieties.[207] In subsequent centuries orange carrots moved wherever Western empires colonized.

Orange carrots were known in Europe from before the Renaissance but they do not appear to have become commonplace.[208] The origin of the orange mutant, and its establishment as the defining feature of a carrot, has provoked extensive academic discussion.[209] Ideas have included direct selection from wild carrots or selection from hybrids between domestic and wild carrots. However, detailed investigation of genetic diversity across many different forms of wild and cultivated yellow, purple and orange carrots has concluded that orange carrots were selected from yellow cultivated carrots.[210]

Carrots have been used since antiquity as medicine as well as food. Carrots are thought to have diuretic effects, although they have also been a popular source of phallic euphemisms, if morphology maketh the man. Associations with fertility and reproduction have ancient roots, as mentioned by Athenaeus. Early modern herbals make frequent reference to carrots as emmenagogues – that is, menstrual regulators – and in the late nineteenth century Scots Gaels in the Western Isles were reported to use wild carrot to symbolize fertility, offspring and children.[211] As carrots are nutritious, easy to grow and also store well, in times of shortage they have become major food sources. During the Second World War the British government's Dig for Victory campaign promoted the cultivation of carrots and emphasized their versatility; it was a time when Britain's population had never been healthier.[212]

Isatis

Waid.

Woad

Isatis tinctoria L.; Brassicaceae

THE BEAUTY OF BLUE

The very mention of woad evokes images of blue-painted Ancient Britons harrying the Roman army. Julius Caesar commented, rather ambiguously, upon blue-stained Britons, while Roman soldiers called northern Britons 'Picts' (Celtic meaning 'painted people').[213] The painted stereotype is so strong that Sellar and Yeatman used it in their humorous account of British history, *1066 and All That* (1930). Whether woad was really used by Ancient Britons to stain their bodies is a source of academic dispute.[214] However, different cultures at different times have used the dye to denote rank, promote privilege, enhance beauty, foster trade, inspire fear and demand respect.

Woad, a chest-height member of the mustard family, appears unremarkable. It has long, thin, blue-green leaves, a strong cabbage-like odour, yellow flowers and flattened lantern-like fruits. However, as the plant ages, a clue to its importance in Western civilization is revealed; it develops a dark blue tinge. Using hit-and-miss chemistry, people learnt to extract a blue dye from woad.

Woad was restricted to neither Britain nor body art. It is native to Central and Western Asia but widely introduced throughout Europe, and was grown and used across Northern Europe as one of the main dyes to colour fabrics. Woad fruits have been found at Iron Age sites across Europe and in ninth-century Viking sites in Britain, although this does not necessarily indicate they used it for dyeing.[215] Combined with other vegetable dyes (red from madder and yellow from weld), European dyers created an extensive palette of coloured yarns. In ancient Rome, woad was a painters' pigment, while at least one artist who contributed to the eighth-century Lindisfarne Gospels appears to have used a woad-based paint.[216]

To use woad as a dye or pigment, the harvested plants had to be treated appropriately; only then would the chemical precursors contained in woad be converted to the blue dye indigotin. The lengthy process was noisome. Traditionally, young woad leaves were harvested, crushed and manually moulded into cricket-ball-sized spheres before being dried and stored. Eventually, the woad balls were pulverized and piles of damp woad powder left to ferment in a process called couching – for weeks piles of fermenting woad powder would release sulphurous gases that smelt of rotting cabbage. During this time bacteria would convert indigotin precursors into indigotin. However, indigotin is insoluble and needed further treatment to make it useful for dyeing fabrics. The indigotin would be dissolved in vats of alkali to make a yellowish dye, indigo white. It was then ready for the dyers to begin work. When fabrics were immersed in the woad vats they emerged yellow-green, but quickly turned blue as they dried. The whole sequence relied on the skills of woad producers and dyers to stop the complex chemical processes at just the right times. Through their practical skills, medieval European dyers, the period's industrial chemists, discovered how to transform and manipulate plant extracts into high-value, portable pigments.

Medieval dyers needed high-quality raw material if they were to produce the best results. Woad was cultivated in areas with highly fertile soils, tended by itinerant farmers. In England, the West Country and East Anglia were important woad centres, while in Europe centres were established in Thuringia, Tuscany, Normandy and Languedoc. The mansions of woad merchants in Toulouse attest to their wealth, as their prime woad-stained cloth was exported to the Low Countries, Britain and Spain.

In modern terms, woad production was unsustainable. For high yields, woad needed rich land and large amounts of fertilizer, and the wealth it promised encouraged farmers to set aside large areas of fertile land to woad – at the expense of food production. Consequently, during periods of political, social and environmental instability the riches of woad production for the few increased the risk of famine for the many. In October 1585 Elizabeth I gave vent to concerns that short-term woad profits were damaging long-term food security, and banned new land being sown with woad.[217] The contemporary English herbalist John Gerard was direct about the impact of woad on people's lives: 'it serveth well to die and colour cloth, profitable to some few, and hurtfull to many.'[218]

Until the eighteenth century, woad was the primary source of indigotin in Western Europe. But it was not the only source. A form of tropical pea, *Indigofera tinctoria*, is a particularly important indigotin source in Eastern and Southern Asia, having been used in India for at least 5,000 years. The tiny quantity of indigo imported to Europe along overland trade routes was a luxury that was no competition to woad producers. However, once the Portuguese established reliable maritime routes to India in the sixteenth century, the competition from indigo became a real threat. Coalitions of woad producers and merchants lobbied governments in England, France and Germany to enact often draconian legislation to prevent dyers from using indigo rather than woad. By the eighteenth century Britain had economic interests in India; indigo cultivation had become vital to the interests of the British Empire and the East India Company. The local interests of British and European woad producers were sacrificed on the ebb and flow of imperial priorities. Ironically, the conflict in Elizabethan England between woad and food production was later paralleled in Victorian India as famine hit indigo-cultivation areas.[219] In North America, hard-wearing, practical, indigotin-stained denim was adopted for work clothes and came to symbolize hard-working pioneers: the birth of blue jeans, although the indigotin came from indigo not woad.

The interests of both woad and indigo producers were usurped in 1883 when the German chemist Adolph von Baeyer discovered how to synthesize indigotin artificially. In 1897 about 19,000 tonnes of plant-derived indigotin were produced annually; a century later a similar amount was produced using Baeyer's industrial process.

Today woad is little more than a cottage industry, despite the wealth and trade routes it generated and the science of chemistry it fostered. The ecological and social realities of woad cultivation and indigotin production belie its modern 'wholesome' image. In some areas, where it was once cosseted, woad has escaped the hand of humans and become a serious agricultural weed.

Citrus

Citrus species; Rutaceae

ORANGES, LEMONS AND LIMEYS

In 1740 a flotilla of British ships, commanded by George Anson and crewed by 1,854 sailors, embarked on a voyage to relieve Spain of her Pacific possessions. When Anson returned to England four years later, having completed a tactically triumphant circumnavigation of the globe, only 188 of his crew were alive.[220] The principal cause of this maritime slaughter was neither shipwreck nor conflict but scurvy. Scurvy blighted maritime exploration, afflicting navies, merchants and pirates alike. Scurvy and the means to prevent it became strategically important issues as European nations began to wield naval power.

Scurvy, caused by vitamin C deficiency, causes lethargy, horrific sores and loss of teeth, jaundice and, ultimately, death. Humans are one of the few mammals that cannot synthesize vitamin C (another, oddly, is the guinea pig), so we must obtain it from our diet. All manner of nostrums were recommended to prevent scurvy; few were effective, many useless and some fatal.[221]

The best way to prevent scurvy is regularly eating fresh fruit and vegetables; items in short supply on long sea voyages. In 1747 an experienced Scottish ship's surgeon, James Lind, tested six widely used scurvy treatments on twelve sailors, in what is acknowledged as the first documented clinical trial.[222] The most successful of these treatments was when sailors consumed two oranges and a lemon daily. Despite resistance from the Admiralty and the conservative eating habits of British sailors, by the end of the eighteenth century citrus juice was part of sailors' rations, and between 1779 and 1813 naval scurvy cases fell by 75 per cent.[223] During the nineteenth century,

a. Malus Aurantia striis argenteis distincta.
b. Malus Aurantia Lusitanica seu Pomum Sinense, Apfel aus Sina.
c. Malus Aurantia folio Salicis, Pomeranze mit Weidenblätter,
d. Malus Aurantia monstrosa foliis et fructu variegatis, Pizarria.
e. Malus Aurantia Indica pumilo dicta, seu humilis.

British sailors, and the British more generally, became colloquially (and perjoratively) known as 'limers' or 'limeys' because the British navy, through confusion over the names of citrus fruits, and to some extent political and economic expediencies, chose to supply their ships with limes rather than lemons. (Limes have only about a quarter of the vitamin C content of lemons, so they were effectively useless, but by then on-board supplies and faster travel times had reduced the risk of contracting scurvy.)

Citrus fruits are technically termed hesperidia, in homage to the belief that the Golden Apples of the Hesperides were either oranges or citrons. We have available to us a large variety of citrus fruits, from the familiar sweet oranges (*C.* × *sinensis*), lemons (*C.* × *limon*) and limes (*C.* × *aurantifolia*) to the large grapefruit (*C.* × *paradisi*) and the little clementine (*C.* × *clementina*), as well as citrons (*C. medica*), pomelos (*C. maxima*) and any number of variations on the neat mandarin (*C. reticulata*). The combination of ready crossing among wild *Citrus* species (hybrids are shown with '×' in their names), effective clonal propagation by seed and millennia of use by humans have produced this diversity in colour, size and flavour, yet as few as four wild *Citrus* species appear to have contributed genes to cultivated citruses.[224]

Citruses appear to have originated in an extensive area bounded by north-east India, south-west China, the Malay Archipelago and Australia, and their slow, global spread from Asia followed trade and colonization routes into Europe and beyond.[225] The northern Indian thick-skinned, aromatic citron was known to classical Greeks and Romans. By the eleventh century, bitter oranges (*C.* × *aurantium*), the main ingredient of marmalade, had been introduced to the Mediterranean, with pomelos and lemons following in the twelfth and thirteenth centuries.

Sweet oranges had been the focus of some of the earliest plant collection expeditions. In Japan, a monument to Tajimamori, who went to China in 61 CE at the command of the Japanese emperor Suinin,

a. Malus citria cornuta fructu magno.
b. Malus Limonia cucumerina, Zucheta, Cucummer „ Limon.
c. Malus Limonia fructu superficie aurantii, Pomme d'Adam, Adams Apfel.

5.

a. Malus sive Poma Adami spinosum.
b. Malus cotonea seu Cydonia, Coignassier, Quitten Apfel.
c. Malus cydonia fructu cornuto, Coignier, gehörnte Quitten.

commemorates the search of the 'ever shining citrus', believed to be an elixir of immortality. However, they were a late addition to the European diet, appearing in the fifteenth century. Once introduced, sweet oranges soon became an important crop in the Mediterranean, and scattered into the Americas by Spanish and Portuguese empire-builders. As a young man, Bernal Díaz marched across Mexico with Hernán Cortés and was involved in the subsequent conquest of Mexico. In his old age he described his role in these events and in an expedition he made to the Yucatan Peninsula from Cuba in 1518. Díaz claimed he had

> sowed these [orange] pips, which I had brought from Cuba, for it is rumoured that we were returning to settle. The trees came up well, for when the *papas* saw that they were different plants from any they knew, they protected them and watered them and kept them free from weeds. All the oranges in the province are descendants of these trees.[226]

In seventeenth-century Europe citruses became fashionable garden plants and a means to display conspicuous wealth, although they were not expected to fruit, as the gardener John Rea wrote:

> The Orenge-tree considered as it groweth with us, may more fitly be placed among the Greens then with the Fruits; for all that the benefit it affordeth us, consisteth in the beauty of the evergreen leaves, and sweet-smelling flowers, the fruit in our cold Countrey never coming to maturity.[227]

Citruses are not frost hardy, so in colder climes they must be moved indoors during the winter. Without a clear understanding of the essential requirements for plants to flourish, citruses were moved into poorly lit heated buildings. However, from these rather mean constructions, orangeries eventually developed, which, with access to the limitless wealth and labour of monarchs, could be magnificent. For the merely very rich, personal orangeries required advances in glass and heating technologies. Methods for directly heating greenhouses, using open fireplaces or pans of burning charcoal, were crude and polluting. With indirect heating, greenhouses became more evenly heated and their atmospheres cleaner. By the end of the seventeenth century English estates could boast productive orangeries, although occasionally at the cost of the rest of the garden. In the glasshouse of Beddington Gardens in Surrey the aged gardener claimed to have 'gathered ... at least ten thousand oranges last year [1690]', although the rest of the garden was 'all out of order'.[228]

Citrus production is no longer the sole domain of wealthy aristocrats; it has become a global industry.[229] In 2022 more than 112 million tonnes of citrus fruits were harvested globally, of which 76 million tonnes were oranges. Brazil was the largest orange grower (nearly 17 million tonnes in 2022), followed by India (over 10 million tonnes). The vast majority of oranges go to make orange juice – still promoted as a valuable source of vitamin C, though the risk of scurvy is rather less these days.

a. Nux moschata, Muscades femelles, Müskat Nüß.
b. Nux moschata mas, Muscades mâles, lange Müskaten

Nutmeg

Myristica fragrans Houtt.; Myristicaceae

THE SPICE THAT SPLIT THE WORLD

A tiny group of tropical volcanic islands were once a most powerful force in the world: they provoked repeated conflict among European nations, they were the reason for the world's first multinational company, and they changed the fate of Manhattan. The Banda Islands, whose total area is about that of Grand Cayman in the Caribbean or half that of the Isle of Wight, are part of the Moluccas or Spice Islands in the South China Sea, and their pivotal historical role arose because until the nineteenth century they were the only places on the planet where the spices nutmeg and mace grew. Europeans played political and economic power games for control of the spice trade, making the Bandanese collateral damage and all but exterminating them.

Nutmeg and mace both derive from the same evergreen tree. Its yellow fruit, the size of a ping-pong ball, has a thick, fleshy wall encasing a hard, shiny brown seed, which is covered in a thin mesh of red, fleshy aril. The dried seed is nutmeg; the dried aril is mace.

Although nutmeg is not essential to human survival, it became valued as both medicine and food. Nutmeg was incorporated into Western pharmacopeias, inevitably (given its exotic origin) as an aphrodisiac and sometimes as a panacea. The Tudor physician Thomas Cogan praised nutmeg as the best food for students, especially 'if they can get nutmeg condite [candied]; ... that they would have alwaies by them halfe a pound or more to take at their pleasure'.[230] Cogan's student audience was evidently wealthy.

Western cultures had no knowledge of nutmeg until about a thousand years ago, when Arab and Persian traders started to import the spice along the Silk Roads.[231] As nutmeg was moved from its eastern source, through multiple intermediaries, to Western consumers, its price, and prestige, increased. Nutmeg became a social barometer; possession made implicit statements about the owner's wealth or status. The rich and well-connected filled their spice cupboards and medicine cabinets with the unusual and the exotic; the rest had to make do with what local economies could afford.

By the eleventh century nutmeg was mentioned in trading documents from Mediterranean markets and on tax, commercial and household accounts across Europe. In England nutmeg was sufficiently familiar for the fourteenth-century poet Geoffrey Chaucer to refer to it, while Shakespeare could be confident his Elizabethan audience would understand its exotic appeal.[232] Yet it is obvious from early modern printed herbals that living nutmeg trees were all but unknown.[233] The most complete first-hand account of nutmeg as a growing tree was made by the German–Dutch botanist Georg Rumphius, the 'blind seer of Ambon', in the 1660s, although his account, following a litany of disasters, was not published until nearly a century later, as *Herbarium Amboinense* (1741).

Since few Western Europeans had travelled to South East Asia before 1500, the only word about the places where nutmeg grew came from Italian travellers such as the Venetians Marco Polo and Nicolò de' Conti and the Bolognese Ludovico di Varthema.[234] Despite the indifference or imprecision these accounts displayed towards precisely locating where nutmeg grew, the search for a western route to eastern spice was one of the drivers for Christopher Columbus's voyage across the Atlantic Ocean.

The situation changed in 1511 when the Portuguese captured Malacca in the Malay Peninsula and sent a small expedition east. A year later the expedition's leader, António de Abreu, returned triumphant, his vessels laden with spices of immense value; he had discovered the Moluccas. Abreu's discovery created a problem for the Portuguese Crown.

In 1494, to accommodate Columbus's discoveries in the western hemisphere and mitigate the effects of an earlier treaty, the Portuguese and Spanish Crowns had drawn up the Treaty of Tordesillas, which placed an arbitrary line from pole to pole through the Atlantic Ocean. Newly discovered lands east of the line would be Portuguese, those to the west Spanish. With the discovery of the Moluccas, the Spanish argued the Treaty of Tordesillas should be extended into the eastern hemisphere, making the Moluccas Spanish. More than two decades of Iberian squabbling ensued until the Treaty of Zaragoza was signed in 1529; Spain relinquished her claims to the Moluccas for 350,000 gold ducats. The changed geopolitical climate meant the Portuguese could establish a risky global monopoly in the nutmeg trade: 'Banda is very unhealthy. Many go there and few come back; yet people are always eager to go there because there is much profit.'[235]

The Treaties of Tordesillas and Zaragoza shaped Iberian global colonization into the twentieth century, but Protestant European powers, keen for their share of riches in the east and west, hardly took notice of the papally sanctioned agreements. In 1602 the Dutch government granted her merchants a monopoly over Asian trade: the Verenigde Oost-Indische Compagnie (VOC) or Dutch East India Company was founded. By the mid-seventeenth century the VOC had ousted the Portuguese from the Moluccas. It asserted its power of life and death over the inhabitants of the East Indies and during the next two centuries became so powerful that it assumed the authority of a state: the world's first multinational company.[236] Some of its most notorious and bloody actions were undertaken during the decade 1619–29 when Jan Pietersz Coen was governor general and the spice monopoly was being established.[237] Moluccans were systematically deprived of land, liberty and life, people were enslaved and nutmeg trees outside selected areas destroyed to control production. Prices were artificially maintained by burning nutmeg stocks, and those who attempted to smuggle seeds or plants were executed. The profits accrued to the VOC during the seventeenth and eighteenth centuries were vast.[238] Like similar colonial activities by other European empire-builders, most direct and indirect beneficiaries of the monopoly ignored the brutalities.

In addition to the VOC's brutal policies, the nutmeg tree itself may have contributed to the company's ability to maintain a production monopoly for so long. The nutmeg's very limited natural range made it easy to restrict access to trees. Furthermore, nutmeg trees are male or female and only start producing fruit after a decade; it takes a long time to discover which trees are going to produce fruit. In addition, the fat-rich seeds cannot be stored for more than a few weeks if they are to germinate, and propagation through cuttings is difficult. The Dutch had to be ever vigilant, though, for nutmegs growing from bird-dispersed seeds: birds eat the aril and discard the seed.[239] The Dutch seem to have used this biological observation to insist that planted nutmeg would not thrive: 'them as Fructifie and arrive at perfection, arise from a ripe Nutmeg swallowed whole by a certain Bird in those Islands, which disgorges it again without digesting it, and this falling to the ground with that slimy matter it brought along with it, takes root and grows a useful Tree.'[240]

The British and Dutch empires consolidated their possessions in the Americas and South East Asia, respectively, at the expense of isolated colonies surrounded by hostile territories. There was an unexpected consequence of Anglo-Dutch tensions in the Moluccas. At the end of the seventeenth century the Dutch agreed to exchange their island of New Amsterdam for the British-controlled Moluccan Island of Run. The spice-rich Run must at the time have seemed the wealthier territory, but times change: we now know New Amsterdam as Manhattan. Ultimately, both empires would lose the economic advantages of these territories by the end of the eighteenth century.

The British repeatedly tried to break the VOC monopoly, eventually succeeding at the end of the eighteenth century, when the gardener Christopher Smith successfully collected nutmeg seed from the

Moluccas and established tens of thousands of trees in Malaysia. (Curiously, Alfred Russel Wallace, the British South East Asian explorer who wrote contemporaneously with Darwin on evolution, strongly supported the Dutch nutmeg monopoly in the Moluccas.[241]) At about the same time, the Frenchman Pierre Poivre smuggled nutmeg plants into Mauritius and established plantations for the French Empire.[242] The French Enlightenment writer Guillaume Raynal compared Poivre's audacious actions to Jason's theft of the Golden Fleece.

Breaking the Dutch nutmeg monopoly deprived the Dutch government of taxes and meant the spice became widespread and cheap. A tree once restricted to a few South East Asian volcanic fragments is today found throughout the tropics, although the vast majority of global nutmeg production (*c.* 134,000 tonnes per year) comes from Indonesia, India and Guatemala. Nutmeg is no longer a mainstay of international trade and has become commonplace, a spice that makes food more interesting. The geopolitical power of nutmeg may have disappeared but the cultural and intellectual achievements of the Dutch Golden Age, financed by the nutmeg trade, remain. Picking through the carcass of the nutmeg trade raises uncomfortable issues about how far businesses may go in the pursuit of profit; issues that businesses of all sorts face today.

White mulberry

Morus alba L.; Moraceae

SILK AND SECRECY

Silk conjures up images of luxury, excess and sensuality, with rulers, governments and prelates encouraging or discouraging its use for economic, social or moral reasons depending on political expediencies. When state-sanctioned English pirates captured the Portuguese treasure ship *Madre de Dios* in 1592 on its return from South East Asia, her fabulous silk-rich cargo was one of the richest ever seized and caused a sensation in England.[243] The cargo conveniently fed popular perceptions of Iberian excess and enhanced the political and economic case that English merchants needed maritime access to East Asia.

The Silk Roads – there was not one single road – are the ancient, often romanticized, overland routes that linked Western Europe and Eastern Asia for thousands of years before maritime routes were established between these two worlds.[244] The routes stretched approximately 10,000 km from Xi'an in central eastern China, across the plains of northern China and the Tien Shan, through the ancient regions of Medea, Persia, Mesopotamia and Cappadocia to the European and North African gateways of Damascus, Constantinople and Alexandria.

Chinese silk is produced by glands in the heads of silk moth caterpillars ('silkworms'). Six weeks after hatching from its egg, the silkworm is ready to spin its snow-white cocoon in which to metamorphose into an adult. From a pair of modified salivary glands on its head it produces two fine streams of protein solution that are bound together by protein-rich 'glue' (sericin) into a single silk filament.[245] The filament is then spun into a cocoon. To harvest the silk thread, farmers kill the

a. Morus alba, Meurier blanc, Weisse Maülbeer. b. Morus nigra,
Meurier, Maülbeer

metamorphosing silk moth and unravel the filament; one cocoon contains up to 1.5 km of silk.

After thousands of years of selection by humans, silk moths became finicky feeders, incapable of spontaneous reproduction and unknown in the wild.[246] They are creations of humans, as dependent on us as the black-and-white cows at the heart of modern industrial milk production. The domesticated silk moth has one food plant: white mulberry.[247] One silkworm consumes approximately 60 g of white mulberry leaves during its six-week life, producing approximately 0.2 g of silk. To put it another way, 1 hectare of white mulberry trees will produce approximately 40 kg of silk.

Mulberries belong to the small genus *Morus*, which is one of a group of genera, including apples, maples and magnolias, whose species are divided between China and eastern North America. These so-called Tertiary relicts are evidence that between 15 to 65 million years ago a vast temperate forest stretched from the Mediterranean through Central Asia across the Bering Straits into North America.[248] White mulberry is the most widespread mulberry species, and is distributed in a fragmented arc that approximates to the Silk Roads. However, the precise distribution of white mulberry is unknown; thousands of years of anthropogenic change have blurred its native range. Two other mulberries have significant supporting roles in the relationship of Western civilizations with silk: the eastern North American red mulberry (*Morus rubra*) and the Southwest Asian black mulberry (*Morus nigra*).

White mulberry is very variable in terms of habit, leaf shape, fruit colour and form, physiology and genetics. Given the vast range of the species and its long history of association with humans, this is hardly surprising. However, white mulberry management has contributed to the variation. As trees are pruned to keep them under control and promote growth of vigorous, nutritious young shoots, mutations can accumulate. Such abnormalities, frequently lethal in animals, are important sources of variation in domesticated plants. New shoots may possess characteristics attractive to a farmer and be propagated to produce orchards of thousands of genetically identical trees. Given the vast quantities of white mulberry leaves needed to feed silkworms, farmers are likely to have unconsciously selected trees with low tannin and high protein contents – they would have observed that silkworms fed on tannin-rich, protein-poor leaves grew slowly and produced little, low-quality silk.

All civilizations need negotiable resources to acquire the things they want, placate their enemies or purchase their allies; they also need resources to satisfy their rulers, pacify their deities and subjugate their populations. Early Chinese civilization was no exception. The 'currency' the Chinese discovered, developed and initially controlled was silk. At least 5,000 years ago the Chinese perfected three essential technologies for high-quality silk production: silkworm growing, silk throwing (making a thread that can be woven from silk filaments) and silk weaving.[249] As an imperial luxury the mysteries of silk manufacture were jealously guarded. However, as demand for silk increased, knowledgeable traders and middlemen became involved and the Chinese monopoly on silk gradually ebbed away.

For at least two and a half millennia, European and North African civilizations imported silk in ignorance of its nature and source. Rome haemorrhaged gold to acquire this seductive, evanescent fabric. At its height, Imperial Rome was importing tonnes of silk annually, sometimes costing its weight in gold, from beyond the eastern limits of its writ in western Turkey. In the first century CE Pliny calculated the Chinese silk trade drained Rome of about 12,500,000 denarii (equivalent to approximately 3.5 tonnes of gold) each year.[250] Emperor Tiberius tried to prohibit men wearing silk, while Seneca and Pliny both complained that silk clothes protected neither the body nor the modesty of the wearer. If 'woven wind' was to be affordable and not to bankrupt Western exchequers, Western populaces had to curb their desires for silk or wrest the secrets of silkworm husbandry (sericulture) from the Chinese.

In 877 CE political instability in China saw the capture of Canton (present-day Guangzhou), centre of the foreign silk trade, and the destruction of its white mulberry trees.[251] This event may have hastened the destruction of the Chinese global silk monopoly, but the secrets of Chinese sericulture had started to leak long before the massacre of Canton. By the second century BCE methods of Chinese silk production were used in Korea, and by 500 CE the technology had reached the Kingdom of Khotan on the southern edge of the Taklamakan Desert. Sericulture gradually spread along the Silk Roads to Constantinople and throughout Asia Minor and Greece, as the vast rewards from silk production became known.[252] Stories about 'the secret of silk' abound. Silk arrived in Khotan, it is told, when a Chinese princess, marrying a prince of Khotan, smuggled silk moth eggs out of China in her hair.[253] Another story of espionage in pursuit of silk relates how the Byzantine emperor Justinian I bribed monks to bring silkworms to Constantinople; the monks eventually returned with silkworm eggs and white mulberry seeds concealed in their canes.

Once the secrets of silk production were out of Chinese hands, silk was made cheaper by making its production more efficient; silkworms could be more productive or silk throwing and weaving industrialized. In the late thirteenth century a silk-throwing machine was invented in Bologna, and the city state became rich, guarding its technological secrets for nearly three centuries. Southern France specialized in silk weaving, the knowledge of which left France in the late seventeenth century as the persecuted Huguenots took refuge in Britain. By the end of the eighteenth century British industrialists, such as George Courtauld and Peter Nouaille, were throwing and weaving silk on an industrial scale. The stage was set for the nineteenth-century British silk industry that, at its height, employed thousands of workers in hundreds of mills. However, despite their industrial flair, the British failed to acquire the skills needed to grow silkworms as the Italians and French had done.

Attempts to farm silkworms in Britain and North America were prolonged disappointments, as they were across all of Northern Europe. With the usual awe at the profits to be made, schemes to introduce sericulture in England and her American colonies started from the early 1600s and continued through subsequent monarchs and presidents into the early nineteenth century; all failed.[254] Unfavourable climates, disease outbreaks and hubris put paid to many schemes, while others foundered on a misunderstanding about silkworm food.[255] The fruits of the *black* mulberry are delicious but the leaves are poor silkworm food. There survive in Britain examples of black mulberries, which are long-lived, from the early seventeenth century, when James I enthusiastically promoted them as silkworm food. In the 1830s US farmers lost fortunes during 'multicaulus mania' (*Morus multicaulis* is a type of white mulberry closely related to *Morus alba*), speculating that sericulture could be established in North America. An unforeseen legacy of mass white mulberry planting has been its establishment as a widespread weed in North America. Despite the failures, sericulture was consistently promoted as a 'means of bettering the condition of the labouring classes'. Confusion over silkworm food continued until the late nineteenth century.[256]

Silk remains a vast industry; in 2021 more than 85,000 tonnes of raw silk was produced and almost entirely fed by white mulberry. Besides the fibre produced indirectly by silkworms, fibres from young stems of white mulberry have been used to make paper. However, white mulberries have also gradually become known in Western cultures for their raspberry-like white to pinkish fruits, which back in the early seventeenth century the plantsman John Parkinson was already extolling as 'of a wonderfull sweetnesse, almost ready to procure loathing when they are thorough ripe'.[257] These fruits, fresh or dried, have been consumed across Central Asia for millennia and, consequently, numerous forms unconsciously selected. White mulberry has also become important as a livestock fodder, perhaps following the example of the protein-rich silkworm diet.[258]

Petum
Tabaccum

1–6. Blüthe
7–9. Frucht
10.11. Saame

Taback.

Tobacco

Nicotiana tabacum L.; Solanaceae

GLOBAL KILLER

Tobacco has become so enmeshed within Western cultures that mythology, memory and marketing would have us believe we have used it since antiquity. In its native South America its properties have indeed been known for some eight millennia, but it was only from the early 1500s that European empires, as they started exploring the Americas, encountered tobacco and were introduced to its use: Western civilizations have been addicted to tobacco for not much more than 500 years.

In the early 1530s the French explorer Jacques Cartier described how the native people of St Lawrence River used tobacco to 'fill their bodies full of smoke, till that it commeth out of their mouth and nostrils, even as out of the Tonnell of a chimney', and was the first European to describe having tried tobacco: 'tried the same smoke, and having put it to our mouthes, it seemed almost as hot as Pepper'.[259] The Spanish playwright and official Indies historian Antonio de Solís y Ribadeneyra was less open-minded, when he described how Montezuma drank chocolate before he smoked tobacco 'perfum'd with liquid Amber … [a] vicious Habit pass'd for a Medicine with the Indians'.[260] But he was writing some 150 years after the events he reported. By the 1560s the French diplomat Jean Nicot had introduced tobacco to the French court; the fashionable elite soon became addicted, the habit diffused across Europe and tobacco consumption went global, diffusing like smoke on a breeze. Tobacco, or drunkwort,[261] was the first botanical luxury to have a global market and become a universal currency.

Common names can be memento mori of extinct civilizations; a transmitted name and a few shards of pot may be all that remains of an entire culture. 'Tobacco' is apparently such a word; it is one of the few words handed down to us from the now-extinct Taino people of the Caribbean.

In the mid-eighteenth century Carolus Linnaeus immortalized Nicot in the plant's generic name, *Nicotiana*, while the addictive chemical, isolated from tobacco in 1828 by the German chemists Wilhelm Posselt and Karl Reimann, was christened 'nicotine'. From the plant's perspective, nicotine is a powerful insect neurotoxin, one of many chemicals it has evolved to deter pests. The genus *Nicotiana* comprises some seventy species, the majority of them from the Americas. The main species used for tobacco production are *Nicotiana tabacum*, known only from cultivation, and *Nicotiana rustica*, native to an area from south-west Ecuador to Bolivia.[262]

An important process in making tobacco palatable is curing the harvested leaves. Curing allows for parts of the chemical cocktail inside the cells of the leaves to be oxidized, degraded and transformed into the flavours and aromas used by tobacco industries to promote their products. Tobacco is used not only for cigarettes, but in cigars, pipes and shisha, snorting as snuff and chewing. With so many different ways to consume tobacco, and its association with vice and pleasure, came endless rituals and pretensions to ensure people remained excluded from particular groups.

Even in the seventeenth century, opinion was not unanimous in support of this fashionable weed. In the early 1600s the regal critic King James I of England, in his famous *A Counterblaste to Tobacco*, condemned tobacco and smoking as

> a custome lothsome to the eye, hatefull to the Nose, harmefull to the braine, dangerous to the Lungs, and in the blacke stinking fume thereof, neerest resembling the horrible Stigian smoke of the pit that is bottomelesse.[263]

Historically, tobacco held the status of a panacea – a universal cure-all – for everything from psychoses to syphilis, but the only universal cure offered by tobacco is death, often slow and lingering. The astronomer and mathematician Thomas Harriot, a member of the first English attempt to settle North America at Roanoke Island (North Carolina) between 1585 and 1587, observed that among the indigenous people tobacco 'preserueth the body from obstructio[n]s ... whereby their bodies are notably preserued in health'.[264] Once back in England, he strongly advocated regular tobacco use and is thought to have died from smoking-induced cancer. It took over two generations after Sir Richard Doll first synthesized the evidence that tobacco is a killer for tobacco companies reluctantly to accept the harmful effects of their wares. In 2004 the World Health Organization estimated that approximately 9 per cent of 58.8 million human deaths were tobacco-related (beaten only by hypertension); since its adoption as a global drug, tobacco has probably killed more of us than any other plant on the planet.[265]

Tobacco is the third most addictive (after heroin and cocaine) and fourteenth most harmful commonly used drug; for comparison, alcohol is slightly less addictive but slightly more harmful.[266] Prior to their encounter with tobacco, Western cultures had experienced few addictive substances; compulsion and addiction were character failures. Gradually there has been the realization that addiction may also be a consequence of the interactions of our biochemistry or physiology with plant products. Individuals are not the only tobacco addicts. Even James I became seduced by the financial rewards that could be made from tobacco, especially as the American colonies started to produce dividends. Countries that benefit from enormous tobacco-based tax revenues or exports are in the invidious position of tobacco dependency, as are many long-term financial instruments invested in tobacco companies.

Globally, between 2007 and 2021 the percentage of adults smoking fell from 22.7 per cent to 17 per cent, although such headline figures mask details. Tobacco production and consumption have declined in developed countries but are increasing in developing countries as tobacco companies refocus their marketing efforts.[267] In 2009 global tobacco production was about 7 million tonnes; in 2022 this had fallen to about 5.8 million tonnes. China, the world's largest producer, supplied about 38 per cent of the world tobacco market, a decrease of about 5 per cent in production compared to 2009.[268] The United States, once famous for the quality of its tobacco, grown and harvested by slaves in the plantations of Virginia, today accounts for approximately 3.5 per cent of global tobacco production.

Yet tobacco is not a 'bad' plant. In the nineteenth century the genus *Nicotiana* became a model for early work on understanding plant hybridization,[269] and towards the end of the twentieth century it became a model to understand fundamental plant biology. In 1982 tobacco became the first artificially genetically modified plant,[270] heralding the genetically modified crop revolution of the early twenty-first century. And even its health dangers have a silver lining: arguments over the toxicity of consuming tobacco have been instrumental in establishing the importance of epidemiology and evidence-based medicine in framing modern health policies.

Tulip

Tulipa species; Liliaceae

FOLLIES OF SPECULATION

In 1841 the Scottish journalist Charles Mackay wove an intriguing account of how, during the seventeenth century, the entire Dutch population succumbed to tulipomania and become obsessed with these colourful flowers.[271] Mackay's interpretation of the events between 1634 and 1637 has coloured Western views of tulips ever since.

In the early seventeenth century the Netherlands were entering the so-called Dutch Golden Age. Enormous wealth from the activities of the Dutch East India Company (see NUTMEG, page 137), the ready availability of highly skilled labour and cheap energy combined to make the Netherlands a European political force and her science and art the envy of Europe. Inside the houses of rich and powerful families, the still-life painting flourished. Outside, Dutch gardeners painted with flowers, creating elaborate gardens filled with exotic plants, accentuating the wealth and social position of their employers.

The Dutch tulip bulb trade, which had started as a spot market involving gardeners interested in tulips, had rapidly evolved into a sophisticated futures market, where transactions no longer needed to involve tulip lovers. Buyers and sellers traded promissory notes, and perhaps never had to see the bulbs at all. The bulbs that attracted the highest prices were those that would produce 'broken' blooms, where the usually solidly coloured tulip tepals were marked with exotic feathered or flamed patterns. Rosen types had red markings on a white background, Violetten were purple-flamed, while Bizarden types had red or purple markings on yellow backgrounds. The most elusive and prized of these tulips was 'Semper Augustus', a Rosen type. Today we know that this unpredictable patterning on some tulips is due to infection by an aphid-transmitted virus.[272] Perhaps there has never been so much money wagered on the outcome of a plant viral infection as there was during tulipomania.

Mackay argued tulipomania was a financial bubble, which continued to expand until someone asked about the 'emperor's new clothes'. This metaphorical question was posed on 3 February 1637, when bulbs failed to sell at a tulip auction in Haarlem. Within days the house of cards that was the Dutch tulip market started to collapse and 1637 became an *annus horribilis* for Dutch horticultural speculators. Buyers owed money they did not have, for bulbs they did not want. Sellers had stock they could not even give away, and were left with debts built on little more than promises. The effects rippled through Dutch society, and even the government got involved trying to restore order.

Over the last thirty years economic historians have challenged Mackay's interpretation of events in the Dutch tulip trade,[273] in particular whether the Dutch tulip phenomenon was a financial bubble and whether its effects rippled much beyond a narrow stratum of Dutch society. Whatever the strict academic merits of the case, tulipomania has become a byword for economic folly, although knowledge of it has not prevented similar and more spectacular financial follies since.

But what of the plant itself that caused all the excitement and despair? There are approximately 120 tulip species, distributed from the Mediterranean and North Africa, through the Levant, Turkey, the Caucasus and Central Asia into China and Japan. The greatest diversity of tulip species occurs in the montane regions of Western and Central Asia, where the winter is very cold, the spring growing season is short and the summer is long and very hot.[274] Botanically, tulips are geophytes, literally 'earth plants', meaning that their underground bulbs are well adapted to survive the winter cold and summer desiccation, and give emerging plants a competitive advantage in spring.

b
a
a. Tu- lipa lutea ſtriis rubris notatis.
b. Tulipa lutea maculis rubentibus.

From wild tulips, the Ottomans and Persians selected and bred the familiar garden tulip, which became a favourite of Ottoman sultans and consequently their courtiers.[275] But we have very little understanding of where, when and how these emblematic flowers originated. They went unrecorded by classical Greek and Roman writers, and when garden tulips arrived in Europe, late compared to other familiar horticultural plants, such as daffodils, roses and stocks, they had already been highly bred and selected. The French naturalist Pierre Belon first drew Europeans' attention to the tulip following his Levantine travels in the late 1540s.[276] In 1554 Ogier Ghiselin de Busbecq, the Flemish-born ambassador of the Holy Roman Emperor to the Ottoman court of Suleiman the Magnificent, reported the prevalence of tulips in and around Constantinople,[277] and just five years later the Swiss botanist Conrad Gesner reported seeing tulips growing in a garden in Augsburg, Germany.[278] The earliest tulip illustration in Western literature is a woodcut in Pietro Andrea Mattioli's *Senensis medici, Commentarii in sex libros* (1565), by which date the Flemish botanist Carolus Clusius was busy distributing tulips around Europe.[279] Clusius, who has a species of tulip named after him, eventually became the founder of the Botanic Garden at Leiden, laying the foundations of Dutch tulip breeding and its current multimillion-euro bulb industry. Seventeenth-century dried tulip specimens in European museums do no justice to the beauty of the living plant; for this one must look to the period's Dutch tulip books, and to the paintings of the era, in which Dutch and Flemish artists caught the flamboyance of these exceptional flowers. Through the paintings of Rembrandt, Bosschaert, van Aelst and their contemporaries we can still feel a little of the excitement that fanned the flames of tulipomania.

Since their introduction to Western Europe, tulips have been popular across the continent. Careful selection has produced hundreds of cultivars of the garden tulip, known as *Tulipa gesneriana* in commemoration of Conrad Gesner. During the eighteenth century and into the early nineteenth century, explorers discovered new species of wild tulip in Europe, Greece and Asia Minor, some of which were brought into cultivation. However, only when the Swiss–Russian botanical explorer Johann Albert von Regel penetrated Central Asia was the full diversity of wild tulips revealed. These wild species provided a new source of genes for tulip breeding. The flame-shaped blooms preferred by the Ottomans have given way to torch-shaped blooms in gardens in the West, and the flower which was once the exclusive preserve of sultans, monarchs and princes of commerce has become commonplace across Western cultures. Annually, during spring binges and optimistic autumnal frenzies, consumers spend vast amounts on tulip flowers and bulbs. And for the tulip connoisseur there is an immense array of different shapes, sizes and colours available, for display and, perhaps, one-upmanship.

Floral frenzies, such as those for auriculas, orchids, ferns, cacti and lilies, have engulfed the horticultural world since tulipomania, although perhaps not to the same extent. However, horticultural obsessions can have destructive environmental consequences. It has become clear that unregulated bulb collecting has had devastating effects on wild tulip populations; some tulip growers are driving to extinction in the wild the plants they claim to love.

a. Solanum mordens bifurcata siliqua. b. Solanum mordens fructu aureo lato. c. Solanum mordens siliquis flavis. d. Solanum seu Piper Indicum maximum. e. Solanum mordens fructu longo erecto. f. Solanum mordens fru= ctu rotundo.

Chilli

Capsicum species; Solanaceae

SOME LIKE IT HOT

The defence chemicals chillies evolved which inflict intense pain on mammals that try to eat them have been subverted by humans; chemical pain has become organoleptic pleasure. Chillies have become culturally important as major elements of cuisines across the world, often far removed from the forests of Central and South America where they were domesticated millennia ago.

The pungency of chillies is due to a specialized metabolite, capsaicin, concentrated in the placenta, those tissues that support the seed inside the fruit. Since the early twentieth century, capsaicin content has been subjectively measured using Scoville heat units (SHU), although today capsaicin content may be measured directly.[280] SHU is a measure of how much a chilli extract must be diluted before its 'heat' is undetectable; the more SHUs the 'hotter' the chilli. Capsaicin contents vary from absent in bell peppers through *jalapeño* (2,500–5,000 SHU) and *habanero* (100,000–350,000 SHU) to extremes such as the 'Trinidad Moruga Scorpion' (2,000,000 SHU).[281] The Naga Hills in India are famous for their fiery chillies such as the 'Naga Viper' (1,400,000 SHU), and young men regularly compete with each other over who can consume the hottest, whilst chilli breeders vie for the accolade of creating the world's hottest chilli. The masochistic consumption of such chillies has been rationalized as the enjoyment of pain without the risk of bodily harm, although more prosaically the ability to win recognition and financial reward may also play a significant role.[282] Biologically, chilli breeding reveals the remarkable amount of natural variability for capsaicin accumulation in wild and cultivated chillies.

Chillies belong to a small genus, *Capsicum*, of approximately twenty-five species that in pre-Columbian times was restricted to the Americas. Five *Capsicum* species were domesticated in the Americas, although the regions of their domestication remain informed speculation: *C. annuum* (bell pepper and *jalapeño*) in Mexico or northern South America; *C. frutescens* (*tabasco*) in the Caribbean; *C. chinense* (*habanero* and Scotch bonnet) in Amazonia (despite its name); *C. baccatum* (*aji*) in Bolivia; and *C. pubescens* (*rocoto*) in the southern Andes.[283] Domestication focused on the fruits; small erect deciduous red fruits of the wild forms were unconsciously selected to produce large hanging persistent multicoloured fruits. Along with these features came changes in capsaicin content and other compounds that give chillies their characteristic flavours. Although all five species contribute to the global spice trade, *C. annuum* is the most important.

Chillies were cultivated and used in cooking across the Americas from the Caribbean to the Andes six millennia ago, although they were probably harvested from the wild at least one millennium before this. As much as 4,000 years ago different chilli species were being cultivated together in Peru, and 3,000 years later multiple cultivars were being grown in pre-Columbian Mexico.[284] By the time Columbus first encountered chillies, they were integral parts of the lives of peoples across the tropical and subtropical Americas.[285]

The first public mention of chillies in Western literature was by Diego Álvarez Chanca, physician on Columbus's second voyage in 1493, under the Taino name *agi*.[286] The name 'chilli' derives from the mainland Nahuatl name that the Spanish later encountered in Mexico in the 1520s. Interest in this new spice established itself rapidly across Europe. The first printed illustrations of chillies were published in the German botanist Leonard Fuchs's *De Historia Stirpium* (1542), and by 1597 the English herbalist John Gerard reported that chillies were 'very well knowne in the shoppes at Billingsgate by the name of Ginnie pepper, where it is usually to be bought'.[287] Unsurprisingly, though,

Gerard was having problems getting the chilli plants he was cultivating in 'hot horsedung' to produce quality fruit because of the 'unkindely yeeres', although he was hopeful of better results 'when God shall sende us a hot and temperate yeere'. Despite chillies being used in 'forren countries' as a substitute for black pepper, and their endorsement by Columbus, Gerard was sceptical since 'it hath in it a malitious qualitie, whereby it is an enimie to the liver & other the entrails'.

By 1615 Francisco Hernández was producing a detailed classification system for the numerous types of chilli being returned to Europe from the New World. Chillies spread rapidly to the Asian possessions of the Spanish and Portuguese empires so that by the end of the sixteenth century they were established in China, India, Indonesia and Japan, whence they started to follow the trade routes from Asia back into Europe.

In his book *Warhaftige Historia und Beschreibung eyner Landschafft der wilden, nacketen, grimmigen Menschfresser Leuthen in der Newenwelt America gelegen* (The true history and description of a country populated by a wild, naked, and savage man-eating people, situated in the New World America) (1557) the sixteenth-century German explorer Hans Staden regales his readers with tales of his three-year incarceration by the Tupinambá people near present-day Ubatuba, in the Brazilian state of São Paulo.[288] Along with details of the Tupinambá's cannabilistic customs and daily lives, Staden describes two types of chilli and the use of chillies as offensive weapons:

> they can ... drive their enemies from the forts with pepper, which grows there. They make great fire when the wind blows, and then they throw thereon a quantity of pepper: if the fumes were to strike into their huts, they would have to evacuate them.[289]

In their warfare, the Tupinambá were taking advantage of capsaicin's ability to irritate mucous membranes – accidentally rubbing your eyes while chopping fresh chillies is something you only do once. Today, synthetic capsaicin is used to similar effect in pepper sprays by security forces and in personal protection devices, and is also exploited in another way. It has long been observed that chilli seeds are dispersed only by birds, as, although both mammals and birds are attracted to chilli fruits, birds do not have the pain receptors that capsaicin targets in mammals.[290] This has led to an effective capsaicin-based rodent deterrent for bird food.[291]

Chillies, along with maize, tomatoes, potatoes and chocolate, are New World food plants that have transformed Western cuisine, often through secondary incorporation of food cultures from Asia such as India and Korea. The defence system that chillies evolved to avoid being eaten has become a case of the biter bit.

Quinine

Cinchona species; Rubiaceae

CURE FOR A COUNTESS

Early apothecaries recognized that, to be effective as a medicine, the correct plant had to be harvested in the right place and prepared in the appropriate manner. A prime example of the importance of these precepts is quinine, traditionally one of the most valuable medicines in Western Europe. Moreover, it arguably facilitated European colonization of the world's subtropical and tropical regions.

In early-seventeenth-century Europe quinine was in great demand as a cure for malaria, and efforts were made to protect the quality of imported quinine bark and the secrets of its preparation. The sale of a quinine bark 'secret' to Louis XIV, in 1679, made the Cambridge quack Richard Talbor a very rich man. Talbor's secret was that he had bought all the best quinine bark in England and ground it in wine to maximize extraction of the active ingredients.

Until it was artificially synthesized in 1945, quinine was extracted from the bark of the quinine tree. The myth of the discovery of quinine is a familiar one. In 1638 the Condesa de Chinchón was dying from an intermittent fever (malaria) in Lima. In desperation a Jesuit priest suggested to the Condesa's husband, the viceroy in Peru, that she might try a local remedy that he had used successfully, although whether the indigenous people of the region knew about the bark as a febrifuge is disputed.[292] The bark proved effective, the Condesa recovered, and following her return to Spain the story became widespread and the rush for quinine bark started. The genus from which the bark was derived, *Cinchona*, was named in her honour by Linnaeus, albeit misspelling her name; an error which, under the rules of botanical nomenclature, has been preserved to the present day.

For over a century little was known about the biology of the tree from which the bark was harvested. Disputes over the efficacy of quinine were associated with the problem that barks from different parts of South America have different properties, while the high price of the bark encouraged adulteration. There was also prejudice over quinine use based on its association with Jesuits and general hostility towards Catholics, especially in Britain and Northern Europe. Indeed, legend has it that Oliver Cromwell, Lord Protector during the English Commonwealth, refused quinine bark as a cure for his (ultimately fatal) malaria because it was a 'Popish remedy'.

The quinine genus is essentially Andean and has a notorious reputation, botanically speaking, with more than 330 names published for the 23 species currently recognized.[293] Until the mid-nineteenth century the only supplies of raw bark were from diverse sources (and hence species) in the New World.[294] Both the British and the Dutch were concerned about bark supply and funded major programmes to collect seed, often illegally, for cultivation in India and Java, respectively. European, and particularly British, justifications for the introduction of quinine bark outside its native range have always been stated in terms of the need to conserve the resource and intervene in the apparently wasteful practices used in the Andes for harvesting *Cinchona*,[295] where mature trees were coppiced or felled to collect the bark. William Dawson Hooker, a son of the soon-to-be director of the Royal Botanic Gardens at Kew, made clear that coppicing was an appropriate method for maintaining productivity and promoting growth of the quinine tree.[296] However, this biological observation seems to have been overlooked by his father when it came to promoting a scheme for introducing quinine to British India.

During the nineteenth century numerous adventurers and scientists went to the New World, either independently or sponsored by

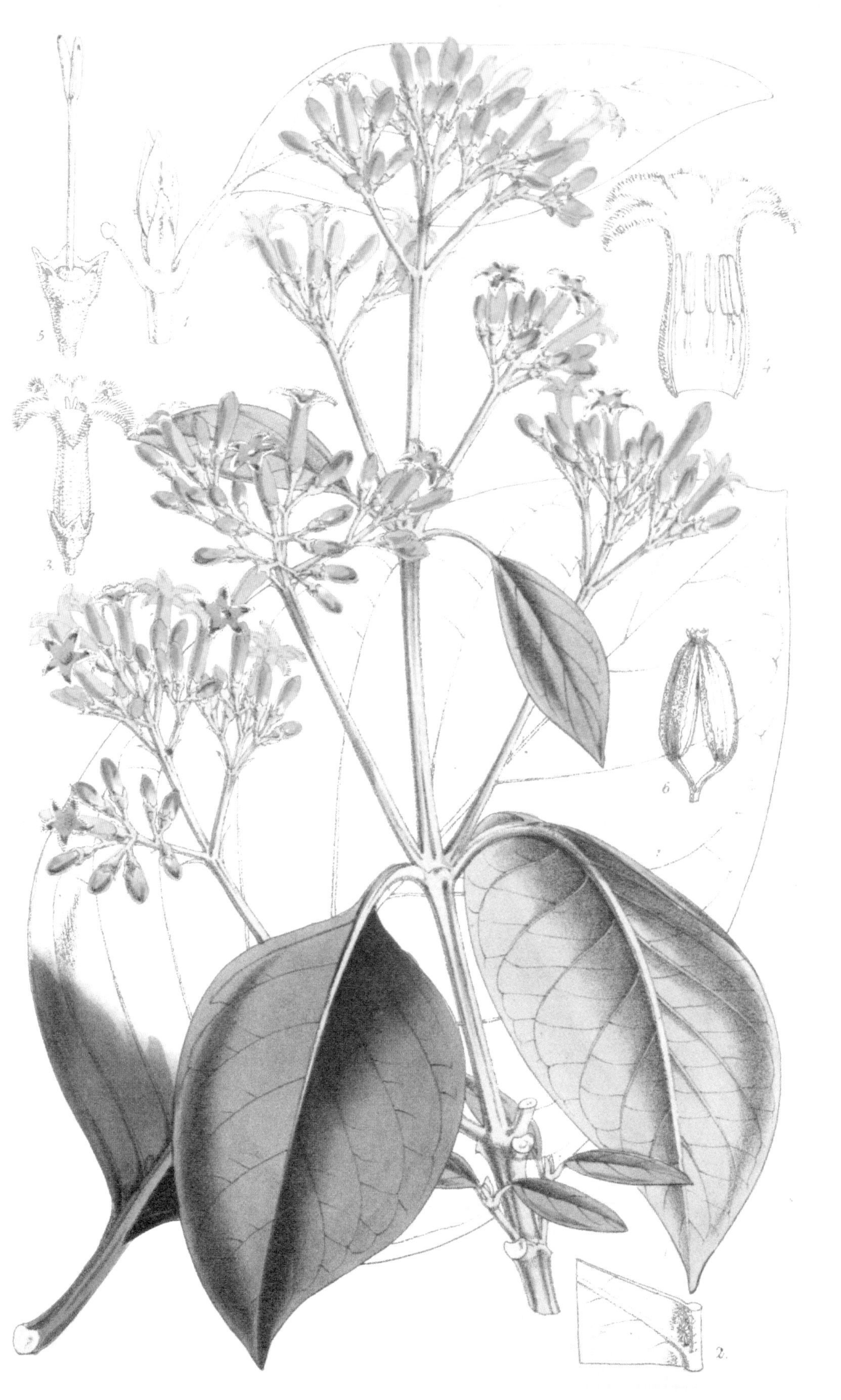

European powers, to search for quinine and bring back material for cultivation. In 1853 the Dutch botanist Justus Charles Hasskarl arrived on the Peruvian–Bolivian border disguised as a German businessman, searching for quinine and illegally procured seeds. In Java, millions of trees were raised from these seeds. In contrast, fewer than 10,000 trees had been raised from Bolivian quinine seeds collected by the botanist Hugues Algernon Weddell two years earlier. The bias towards trees raised from Hasskarl's seeds was a costly error, as the trees raised from Weddell's seeds produced about twelve times as much active ingredient as those produced from Hasskarl's.

In 1860 the Amazonian explorer Richard Spruce collected seeds and plants of quinine trees from Ecuador at the behest of the civil servant Charles Markham, after securing the rights to collect the seeds from the local landowner.[297] Trees from this material were eventually established in India and, until the early twentieth century, were the primary quinine source for British India. The other major source of quinine trees was the businessman Charles Ledger, who in 1865 obtained a 40-pound bag of quinine seeds from Bolivia. The British turned them down as they had their own material growing in India. The Dutch grudgingly took 1 pound of the seed and transported it to Java, where they found that it contained up to thirty-two times the active ingredients of their existing plantations. Ledger had illegally collected his material, since the export of quinine was a Bolivian government monopoly. Similarly, efforts by the British, after Spruce's successful expedition, to collect in other parts of Ecuador were not encumbered by the knowledge that material was being illegally collected:

> It was stated to me that the Government of Ecuador has passed an edict prohibiting the exportation of either seeds or plants of the quina tree, under the penalty of 100 dollars for every plant, and for every drachm of seed. However, after consulting with Mr Mocatta [British vice consul], I undertook to go to Loxa and make a collection of seeds of the *C.* Condaminea.[298]

Quinine bark shows great differences in the levels of active compounds not only across the range of the genus, but within populations of a species and even within the same tree. Such differences may have been responsible for the variable efficacy of quinine bark when it was first introduced to Europe.

Success with quinine by the British and the Dutch in their respective colonies illustrates the importance of variation, knowledge and supply in relation to plants. Ledger, sampling quinine seeds in Bolivia, used the knowledge of the indigenous Bolivian Manuel Incra Mamani, who was beaten to death in 1877, apparently for his part in Ledger's illegal venture. These seeds were grown in Javanese plantations, where access to the supply could be maintained, and formed the basis of the Dutch quinine monopoly right up until 1939. In the post-war years synthetic anti-malarials have become important for the prevention and cure of malaria, and the value of quinine bark extracts has diminished – its most familiar use nowadays is as the bitter flavouring in tonic water.

Cacao
Cacavate.

1.2. Blüthe
3.4. Frucht
5–8. Saame

Cocoa

Theobroma cacao L.; Malvaceae

FOOD OF THE GODS

In its native range in Central America cocoa was combined with water, maize and chillies to make a bitter drink associated with religious and regal rituals. In European hands it was transformed with milk and sugar to create a drink that was a status symbol: chocolate. It also gave Europeans their first introduction to caffeine.

Cocoa is one of a handful of crops (including maize, tomato, chilli and vanilla) originally domesticated in Central America and transported to Europe by the early Hispanic conquerors of the New World.

Cocoa is a small evergreen tree that, unusually, produces clusters of flowers directly on its trunk. Mature trees can produce thousands of small pink midge-pollinated flowers, but only a few ripen into fruit. Each fruit is a slightly ridged yellow–orange melon-shaped pod that has a hard shell and is about the length of a person's forearm. When ripe, the pods are packed with seeds (cocoa beans) surrounded by sweet, fleshy, white pulp. The cocoa beans are rich in fat, to provide energy for the germinating seed; as this quickly becomes rancid, the seeds germinate soon after they are dispersed. Their fat, cocoa butter, is popular as a body lotion, but it is the alkaloids theobromine and caffeine in the beans, part of the plant's chemical defence system against insect and fungal attack, which are the reason cocoa is so sought after worldwide.[299] They are what make chocolate so desirable.

There are about twenty species of cocoa, all native to the understoreys of humid tropical forests through Mesoamerica and northern South America, but only one is the source of commercial cocoa. Approximately ten pods are needed to produce 1 kg of raw cocoa.

Linnaeus christened the cocoa genus *Theobroma* or 'food of the gods', which was appropriate since one Aztec legend has the deity Quetzalcoatl being punished for giving the secret of cocoa cultivation to humans. Linnaeus would not use the common name chosen by earlier botanists for cocoa, *cacahuatl*, since he thought it 'barbarous'.

Theobromine is rare in the plant kingdom: in Mesoamerica it is only found in the genus *Theobroma*. Consequently, archaeologists can use theobromine as a 'fingerprint' for cocoa culture, especially since minute quantities of theobromine can be detected using modern analytical techniques. Potsherds at least 3,500 years old, from a Mayan site in Honduras, show traces of theobromine, the oldest evidence for the use of cocoa. Furthermore, the evidence suggests these people may have been fermenting the fruit pulp to make beer.[300] Theobromine traces on New Mexican pottery suggest North American people had contact with Mesoamerican cocoa cultivators at least 500 years before the arrival of the Spanish.[301] The high value of cocoa in Mesoamerican cultures meant complex trade networks developed. For example, the Aztecs lived in areas unsuitable for cocoa cultivation but exacted tribute from surrounding nations in the form of cocoa beans and used them as 'coins' in everyday transactions; money literally grew on trees. Mayans probably planted cocoa plantations near their homes, presumably having collected seeds from nearby forests.[302]

Genetic analyses of samples collected across the cocoa tree's range indicate that the western Amazon, more specifically the Peruvian–Brazilian border, is likely to have been where humans first domesticated cocoa.[303] Researchers have speculated that initial domestication was for the sweet pulp surrounding the seeds, and that cultivating cocoa for the beans followed the plant's transport to Central America.

When Columbus captured a Mayan trading canoe off the coast of Honduras in August 1502, the cocoa beans he found made little impression upon him. However, the Spanish conquistadors began to export cocoa beans to Spain soon after. Over the next two centuries the popularity of cocoa gradually spread through Spanish society and across Europe, although early views of this new drink were mixed: 'it seemed more a drink for pigs, than a drink for humanity ... the taste is somewhat bitter, it satisfies and refreshes the body, but does not inebriate', wrote one sixteenth-century historian.[304] Western cultures were benefiting from two millennia of Mayan horticultural skills but, as the European chocolate market developed, supply could only be met by exploiting cheap labour – slaves; Mesoamerican Amerindians and Africans, and then, in the twentieth century, indentured labour. As with many consumer products produced in developing countries, the social and environmental ethics of cocoa production have become a twenty-first-century issue for Western chocolate lovers.

There are three main types of cocoa. The high-quality, low-yielding *criollo* types come from Central America, while high-yielding, hardy and vigorous *forastero* types come from the Amazon basin. The Trinidadian *trinitario* type is a hybrid between the *criollo* and *forastero* types and combines the characteristics of both parents. In the 1820s the Portuguese transported *forastero* seedlings from Brazil to West Africa, while other colonial powers hastened to establish trees in their own tropical possessions, the British in Ceylon (Sri Lanka) and the Dutch in Java. Today, 80 per cent of the world's cocoa production is from West African *forastero* types; hardiness has won over flavour.

The manufacture of high-quality chocolate requires attention to detail. Cocoa beans must be harvested when the seeds contain maximum cocoa butter content and the pulp has maximum sugar content. This ensures optimal fermentation to convert chemicals in the seeds to the familiar chocolate flavours. The fermented beans are then dried, cleaned, roasted and ground to produce cocoa mass, which may be further processed into cocoa solids and cocoa butter. Manipulating proportions of cocoa solids, cocoa butter, milk and sugar produces different types and qualities of chocolate.

Nineteenth-century industrialists who made advances in the manufacture of chocolate are household names, including Nestlé and Hershey. In Britain, Quaker families, including Cadbury, Fry and Rowntree, were involved in chocolate manufacture by the 1760s. True to their convictions, these firms were seen as model employers, but by the end of the nineteenth century Quaker principles in chocolate manufacture had melted away. Industrialization democratized cocoa consumption, understanding cocoa chemistry helped create blocks of solid chocolate, while sophisticated advertising made chocolate something people craved. For some, a dispensable luxury had become an indispensable staple.

During its four-century transformation from divine drink to irresistible confection, cocoa has been invested with all manner of properties: mood changer and stimulant, improver of cardiovascular function, through to poison and cause of childhood obesity. Yet it is unlikely to lose its status as one of the world's most tempting treats.

Potato

Solanum tuberosum L.; Solanaceae

FOOD AND FAMINE

Foods such as potatoes, chillies and tomatoes have become so commonplace that we can hardly conceive of a time when they were not part of our cooking repertoire. However, all these plants are part of what American historian Alfred Crosby termed the 'Columbian Exchange', and were unknown in Europe before 1492. We like foods with which we grew up and are familiar; we are inclined to resist changes in our diets, especially if told the change might be good for us. But potatoes sustained the Incan Empire for nearly four centuries before the arrival of Europeans,[305] and by the mid-eighteenth century potatoes had been accepted by Europeans as a 'wonder food'. And so they were: they were cheap, filled people up and did not spoil readily. Potatoes proved versatile too: they could be boiled, steamed, baked, fried or dried. Today, potatoes are the prime carbohydrate in the diets of hundreds of millions of people. Furthermore, potatoes are important in the manufacture of spirits such as vodka and poteen, and as animal food, while potato starch is essential for the industrial production of processed food, adhesives and paper.

Solanum is one of the world's largest plant genera. Like the carrot family, the potato family has both toxic and non-toxic members. Mandrake and tobacco kill or cure depending on dose, whilst potatoes and tomatoes are important foods. As part of their chemical defence system, most *Solanum* species contain a toxic glycoalkaloid called solanine. Potatoes are no exception, and the green parts of a potato plant contain the poisonous solanine. You will probably have noticed that when potatoes are stored in the light, they begin to turn green: this is

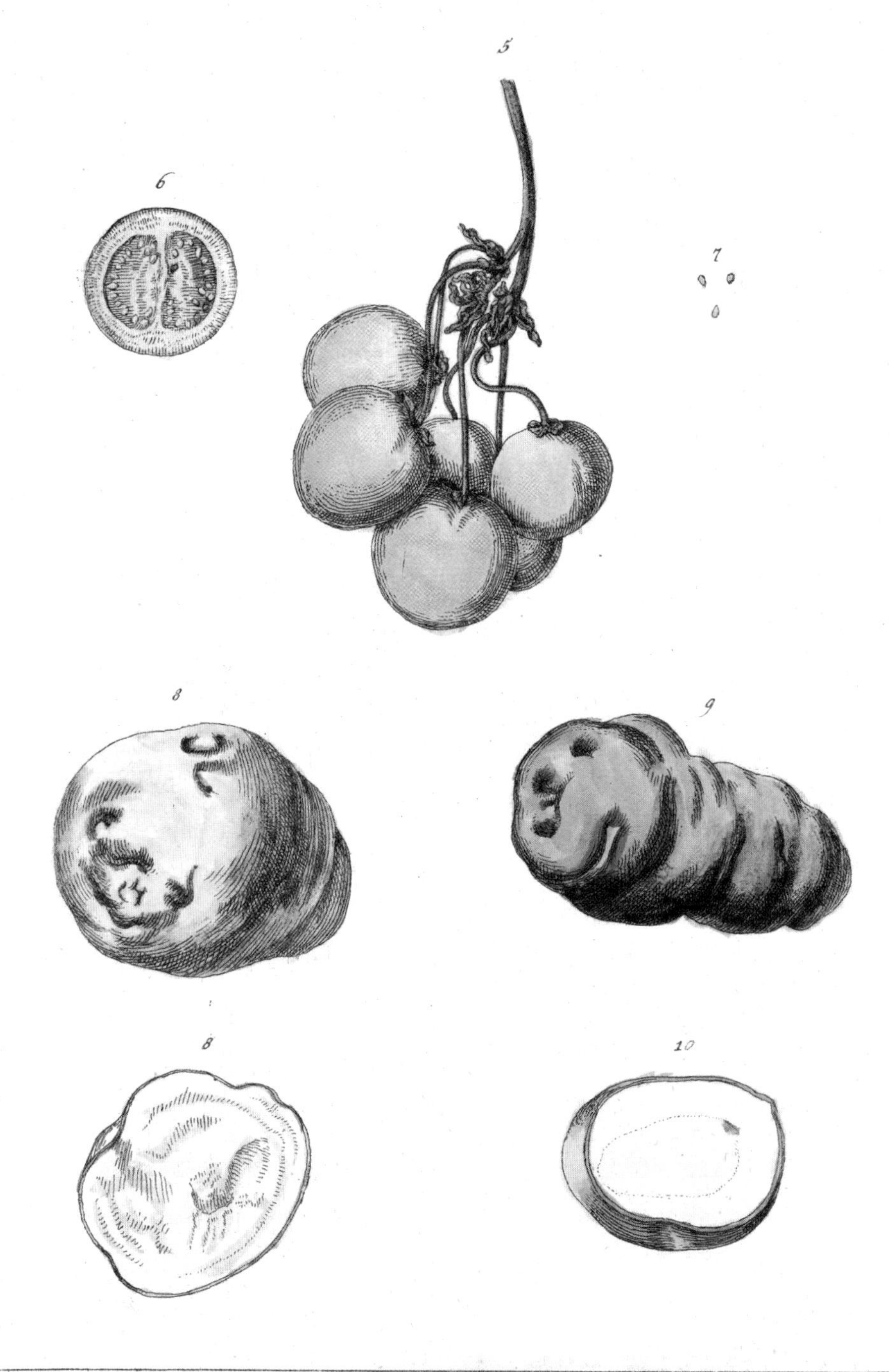

Solanum tuberosum.

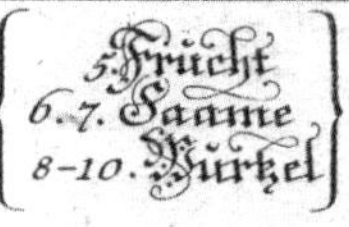

Erdbirn.

indicating that solanine levels are starting to build up, a defence against herbivores and pathogens.

Solanum tuberosum appears to have been domesticated once, more than 5,000 years ago, from a wild species in a region of southern Peru and north-western Bolivia.[306] Millennia of selection by indigenous peoples transformed the toxic tubers of the wild species into thousands of appetizing potato cultivars varying in shape, colour, texture, cooking properties and growth characteristics. Evidence for the central role of potatoes in the lives of pre-Columbian Andean peoples is seen in the form and decoration of their pottery.[307]

Across South America the domesticated potato has been split into two major subspecies: ssp. *tuberosum* from the lowlands of south-central Chile and ssp. *andigena* from the highlands of the Andes. Both subspecies may be distinguished morphologically, anatomically, physiologically and genetically. For example, Chilean potatoes produce their starchy tubers in response to long days, while Andean potatoes produce theirs in response to *short* days.

Outside South America, potatoes first made their appearance in the Canary Islands in 1567.[308] Some Europeans were quick to adopt these curious new vegetables – the Prussians, for example, were planting them in the early seventeenth century, and by the eighteenth century had a thriving potato economy. The English herbalist John Gerard was a potato enthusiast: 'equall in goodnesse and wholesomnesse ..., being either rosted in the embers, or boiled and eaten with oil, vinegar and pepper, or dressed any other way by the hand of some cunning in cookerie'.[309] Others, such as the French and Italians, were wary and needed more persuasion.

French potato culture is inextricably linked with the persuasive power and personality of the eighteenth-century French chemist Antoine-Augustin Parmentier. During the Seven Years War Parmentier was held prisoner by the Prussians and learnt that potatoes had food value. By the 1770s, following his release, Parmentier had scientific evidence of the nutritional qualities of potatoes. Through a combination of panache, guile and luck, Parmentier succeeded in changing the eating habits of a nation. He used special dinners to feed potatoes to 'celebrities' from both factions of the imminent revolution and even presented Louis XVI and Marie Antoinette with a bouquet of potato flowers. On a piece of land in a poor part of Paris, Parmentier grew potatoes; by day, the crop was well guarded but at night the guards left and, predictably and as Parmentier intended, the local populace stole the produce and developed a taste for potatoes. His luck came with the poor cereal harvests of the 1780s and the revolutionary turmoil of the 1790s; starvation finally convinced the French to take potato eating seriously. Parmentier survived the French Revolution and was made a member of the Légion d'Honneur by Napoleon.

John Christopher, in his novel *The Death of Grass* (1956), subtly conceived a world where a virus laid waste all the world's grasses. As cereals and pasture lands disappeared, society collapsed, people faced starvation and the British government detonated a nuclear warhead to reduce the country's population. In Christopher's world, the remnant British population survived on potatoes. However, the precarious state of a human population sustained by potatoes had been shown a century before Christopher's imagined catastrophe.

When the potato fungal infection late blight arrived in Ireland in the mid-nineteenth century it caused devastation. The fungus had been accidentally introduced from the Americas to Europe in the early nineteenth century.[310] Although potato crops failed across Europe, the effects in Ireland were magnified through the mass dependency of the Irish poor on a single potato variety, the 'Irish Lumper', and a complex nexus of political, religious, social and economic issues. The Great Famine that ensued (1845–52) had dramatic effects that reverberate to the present day; mass starvation killed or forced the emigration of up to one-quarter of the Irish population.[311]

The Great Famine and the growing dependence of European economies on potatoes placed a premium on understanding the botanical origins of this commonplace vegetable. The precise South American origin of European potatoes has been debated for decades. Two hypotheses have been favoured. The Chilean hypothesis posits that European potatoes were introduced from south-central Chile. A more recent hypothesis, which has prevailed over the last sixty years, has been the Andean hypothesis, where European potatoes were introduced from the northern Andes; the earliest known herbarium specimens of European potatoes are of the Andean type.[312] The predominance of the Chilean type in modern cultivars is thought to be a consequence of attempts to breed resistance to late blight into potatoes after the Great Famine. Interspecific hybrids with Mexican wild potato species have also been an important source of genes for resistance to late blight.

Recent DNA analyses, however, reveal extant Canarian potato varieties are a mixture of Chilean and Andean types.[313] This observation is important because the Canary Islands were the historical gateway of potatoes into Europe. Using DNA extracted from European potato specimens preserved in herbaria it has been shown that the Andean type appeared in Europe around 1700 and persisted until the end of the nineteenth century, long after the late blight epidemic was over. The Chilean type first appeared in Europe in 1811, long before the late blight epidemics, and persisted until the present day.[314] Together these data support the earlier Chilean hypothesis, emphasizing that the first potato introductions were pre-adapted to European day lengths.

The demands of industrial agriculture have produced highly bred, high-yielding potatoes that have been introduced back into Andean agricultural systems. Understandably, people have chosen these new potatoes in preference to varieties that generations of Amerindian communities have bred, which are now in danger of disappearing. These old Andean varieties must not be scorned or lost, however, as, ironically, it is they that are the reservoir of genetic variation needed for future potato breeding.

The conquerors of the Andean cultures set off from Europe with the dream of discovering El Dorado and untold wealth in gold and silver; what they found proved, in the long term, more valuable than the immediate short-term rewards of precious metals.

Tomato

Solanum lycopersicum L.; Solanaceae

LOVE APPLES TO PIZZAS

Tomayto, tomahto... the world is divided in two about tomatoes, but not just in the way the Gershwin song suggests.[315] To the botanist a tomato is a fruit, but to the cook it is a vegetable; no amount of argument will reconcile the two sides. However, tomatoes also unite people, especially through food, and they are essential in popular Western foods such as pizza, baked beans and ketchup. The tomato is not only a ubiquitous food; it is symbolic of how humans have colonized the planet. Tomato seeds, which evolved to be dispersed by animals, find temporary residence in our guts. Eventually, tomato plants may become weeds wherever we choose to unload our bowels.

Yet tomatoes have been part of European life for fewer than 500 years and widely available for fewer than 150 years.[316] The rise of tomatoes as tasty ingredients with multiple uses in the European kitchen was preceded by deep suspicions they were poisonous or at best fit only to be botanical ornaments. Towards the end of the twentieth century tomatoes were once again the cause of suspicion, as sceptical European consumers became concerned with revolutionary plant-breeding technologies.

Tomato classification is controversial.[317] Linnaeus included tomatoes in the very large genus *Solanum*, while the English botanist Philip Miller separated tomatoes into the genus *Lycopersicon*, a small group of species from the south-western Andes and Galapagos Islands. Based on DNA analyses, current opinion favours their inclusion in *Solanum*.[318] The similarity of wild tomato species and the ease with which they interbreed among themselves and with cultivated tomatoes mean trying to understand the evolutionary origin of the domesticated tomato is a complex task. However, the economic value of the tomato and the desire of breeders to investigate the genetic bases of production and yield characteristics such as disease resistance mean sophisticated genetic analyses have been applied to the problem. Such studies reveal that the tomato as we know it was probably domesticated in two phases. The first phase appears to have occurred in Ecuador and northern Peru and involved the transformation of the wild, tiny-fruited (redcurrant-sized) *Solanum pimpinellifolium* into weedy *Solanum lycopersicum* var. *cerasiforme* with fruit somewhat akin to the cherry tomato. During a second Central American phase of domestication, this spindly vine became the tomato with which we are familiar today,[319] some bearing individual fruits up to 1 kg in weight.

We do not know exactly when the Spanish brought the tomato to Europe, but by 1544 the physician Pietro Andrea Mattioli reported a plant with edible red and yellow aubergine-like fruits was being grown in Italy. These were eventually christened *pomi d'oro*, 'golden apples' (hence the modern Italian name, *pomodoro*),[320] and by the end of the century the English herbalist John Gerard was referring to them as 'apple of Love' or 'Poma Amoris',[321] and was able to report a range of names being used in Europe based on 'love apples' or 'golden apples'. Classical names, such as 'lycopersicon' (wolf peach), were also becoming associated with the tomato. The name 'tomato', derived from the Nahuatl word *tomatl*, only started to appear in English at the beginning of the seventeenth century and did not become commonplace until the eighteenth century.

The herbalist John Parkinson reported that the English 'onely have them [tomatoes] for curiosity ... and for the amorous aspect or beauty of the fruit'.[322] Although by 1648 they were being grown in Oxford Physic Garden, labelled as 'Great Apples of Love', 'Middlesiz'd Love Apples' and 'Little Love Apples' and a decade later differentiated into yellow- and red-fruited types,[323] the British remained cautious about

a. Solanum pomiferum seu Poma amoris, Pomme d'Amour, Liebes-Äpffel.
b. Solanum pomiferum fructu rubro majus, grosse Liebes-Äpffel.
c. Solanum seu Poma amoris fructu luteo.
d. Solanum pomiferum, fructu rubro minore.
e. Solanum pomiferum seu Poma amoris fructu luteo minore, Pomme d'Or, Gold-Äpffel.

these novel fruits. Despite reports that tomatoes were eaten in 'Spaine and those hot regions', the English thought they 'yeelde very little nourishment to the bodie',[324] although they might be 'good to cure the itch' (syphilis).[325] By the nineteenth century, however, the tomato was well established as a high-quality foodstuff across European empires.

Tomato plants are sticky, herbaceous, yellow-flowered scramblers with a characteristic scent, which while spicy in low concentrations may become nauseating in the confines of a commercial glasshouse. A characteristic that distinguishes wild and domesticated tomatoes is that wild tomatoes are not self-fertile since the male and female parts are physically separated from each other. In commercial tomatoes, the male flower parts tightly surround the female parts, making the flowers self-fertile. However, pollen must be physically transferred from the male to the female parts. Outdoors, this is done by bees vibrating their wings to dislodge pollen; indoors, tomato flowers must be mechanically pollinated.

Thousands of tomato cultivars have been bred for growing under diverse conditions and for diverse uses, including pulping, drying and juicing. Frequently grown tomatoes in nineteenth-century Europe rejoiced in names such as 'Mammoth tomato', 'Trophy tomato', 'Apple-shaped purple tomato' and 'King Humbert tomato'.[326] Today, one of the most popular tomatoes among gardeners is aptly named 'Moneymaker'. Despite the very varied appearances of different tomato cultivars, as few as four genes appear to be responsible for the eight broad classes of tomato shape, which range from the flattened beefsteak through rectangular plum and familiar round types to heart- and pear-shaped types. Tomatoes are usually thought of as being red, although they can be everything from orange and yellow to dark purple, with individual fruits varying greatly in size from 5 mm to 10 cm in diameter.[327] International trade has meant fresh tomatoes, once a seasonal treat, are now available year-round. Despite the diversity of available tomatoes, some supermarket tomatoes have reputations for being insipid, with the consistency of snooker balls. Such products are a consequence of having to transport delicate fruits over long distances, present them attractively to demanding customers, whilst maintaining long shelf and fridge lives.

The desire to improve flavour in modern tomatoes led to the development of the first genetically modified plant endorsed for human consumption. The awkwardly named 'Flavr Saver' tomato was genetically modified to slow the ripening process while retaining its natural colour and flavour. Released in the USA in 1994, the tomato was withdrawn from commercial sale three years later.[328] At a similar time, in the United Kingdom, a separate genetic modification was incorporated into a tomato used for paste production. The paste was commercially successful but by the end of the decade mistrust over genetically modified food was widespread in Britain and Europe, so the product was withdrawn. However, the potential health benefits of plants with specific genetic characteristics may yet sway public opinion. If antioxidant-rich tomato becomes the healthy option, 'tomato' may become synonymous not with red but with purple, as heightened anthocyanin levels, which turn fruits and vegetables purple, appear to be key to improving dietary and keeping qualities.[329]

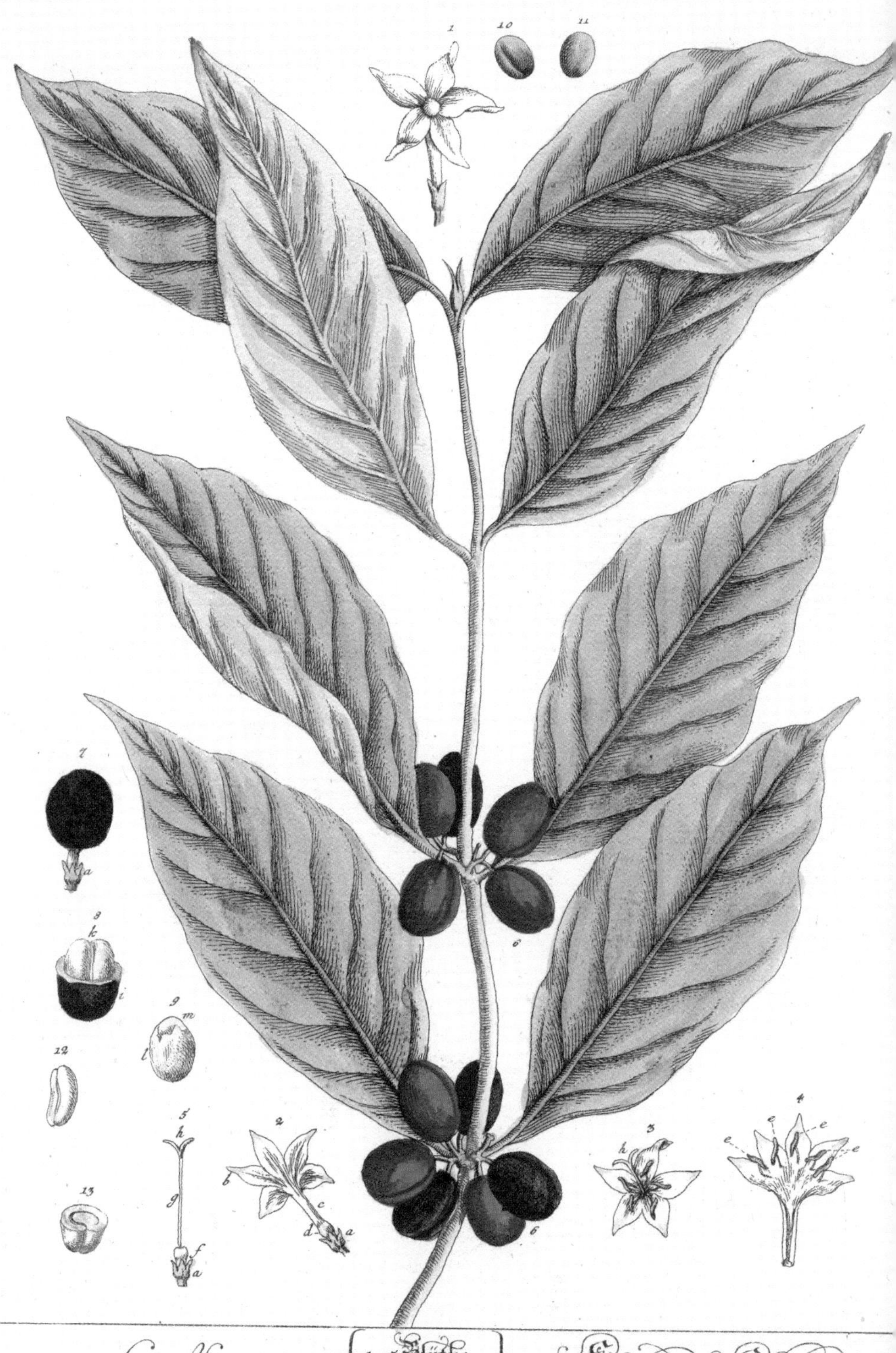

Coffee.

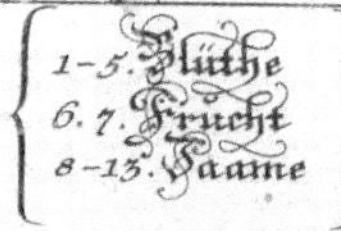

Coffee

Coffea arabica L.; Rubiaceae

ETHIOPIAN SHRUB TO GLOBAL DRUG

In eighteenth-century London, James 'Don Saltero' Salter's celebrated coffee house in Chelsea was a popular attraction.[330] Salter's was a 'cabinet of curiosities' populated by cast-offs from Hans Sloane, whose own collection founded the British Museum, and apparently from donations by notables including Sir Francis Drake and Tobias Smollett. Glass cases stuffed with hundreds of objects competed for space with items such as the 'pizzle of a whale', the 'dried cat' and the 'penguin from Falkner' islands' hanging from the walls and ceiling.[331] As a coffee house, Salter's establishment was perhaps rather unusual, but it was trying to compete with hundreds of other London coffee houses established following the introduction of coffee to Britain in the mid-seventeenth century.

Today, caffeine is one of the most widely consumed legal drugs in the world and is a global addiction; the equivalent of 100,000 tonnes of pure caffeine is consumed annually. For plants, caffeine and its derivatives are biochemical weapons, helping to protect them from disease and insect damage. For humans, these compounds stimulate the central nervous system, relax smooth muscles and have diuretic effects. We have discovered numerous sources of caffeine in plants, ranging from Chinese tea and Central American chocolate through West African cola, South American maté and Amazonian guaraná to Ethiopian coffee.

The coffee genus comprises some 100 species, distributed through tropical Africa to the Mascarene islands of the Indian Ocean, but only three species are commercially significant. The most important, by far, is *Coffea arabica*, an evergreen, fragrant-flowered, red-berried shrub from the mountains of south-west Ethiopia. Tradition would have us believe a ninth-century Ethiopian goatherd discovered coffee when he noticed how giddy his goats got after eating coffee berries. Whatever the means by which the properties of the roasted coffee seeds were discovered, by the sixteenth century coffee was in use across the Islamic world and spreading into the eastern parts of the Christian world.

With the assurance of Pope Clement VIII that coffee was a Christian drink, coffee houses started to appear in Europe from 1645. Public coffee houses in Britain began to spread about five years later, and over the course of the eighteenth century became places where social classes mixed, business was conducted and new, often radical, ideas – political, economic and scientific – discussed.[332] From the ferment of caffeine-fuelled social and intellectual debate, two long-lasting institutions emerged in London, the Royal Society and the insurance marketplace Lloyd's.

Rulers in both the Islamic and the Christian worlds initially treated coffee houses with suspicion; in 1675 Charles II of England, for example, issued a short-lived proclamation against the coffee houses, calling them 'seminaries of sedition'.[333] But there was no stopping the demand for coffee, and states across Europe were unwilling to have supply controlled by the Islamic world. Consequently, during the early eighteenth century, the Dutch, French and British made great efforts to establish coffee plantations in their overseas possessions. In 1718 a Cambridge professor of botany and horticultural enthusiast, Richard Bradley, was eloquent about the Dutch and their success with growing coffee, not just in their warmer colonies but in the Netherlands itself. Bradley was convinced coffee production 'in the *South* Parts of *Carolina*, … would be well worth our Trial, if that Country remains in our Hands'.[334] However, coffee failed to become a significant crop or popular drink in colonial America. During much of the eighteenth century most of the world's coffee was being exported from the Caribbean. Furthermore, wrote Benjamin Moseley,

> the cultivation of coffee requiring but little capital, is an inducement for people of small fortunes to settle in the Islands [Jamaica]. It is a creditable refuge for the industrious man, who has been unfortunate in Trade, and to those whose larger schemes in life have failed. – It is easy employment; the labour light, and many parts of it performed by children.[335]

Moseley's words mask the reality of life for the slaves who worked the plantations that produced the beans to make the 'intellectual drink' of European coffee houses.[336] Conditions on coffee plantations were one of the causes of the Santo Domingo revolution at the end of the eighteenth century, which led to the foundation of modern Haiti.[337]

During the nineteenth century Brazil razed her coastal forests, especially in the south-east of the country around Rio de Janeiro and São Paulo, and planted coffee; the slave-dependent Brazilian coffee boom had started. By the end of the century Brazil's economy was dependent on coffee exports, but slavery had been banned and the plantation owners were reliant on indentured and immigrant labour. The crash in coffee prices during the 1930s had predictable damaging effects on the Brazilian economy and the lives of her people: at the peak of the coffee boom Brazil was supplying nearly 40 per cent of global coffee needs. Despite the bust of the early twentieth century, in 2022 Brazilian farmers grew approximately 30 per cent of the 10.8 million tonnes of coffee produced globally. In the same year the world's greatest importers of coffee were France and the USA.

Coffee is a high-status, high-value product often promoted for its medicinal, taste, stimulant or even temperance values.[338] Despite Moseley's eighteenth-century optimism that coffee was 'incapable of adulteration', by the middle of the nineteenth century coffee-planter Peter Simmonds was complaining that 'no Article ... presents greater facility for adulteration than coffee'.[339] The value and status of the product made it too tempting for the criminal to ignore, and Simmonds went on to enumerate the activities of 'liver-bakers', who cooked and ground ox and horse livers to resemble ground coffee, and described substitutes for ground coffee that included ground acorns, peas, beans, red clay and mahogany sawdust. In other cases, the benefits of fake coffee were promoted – for example, the entirely unappetizing diet drink 'Kalopino' or concoctions based on chicory or roasted barley.[340]

Competing coffee shops now jostle for customers along many shopping streets in Western countries, just as they did over 200 years ago. All manner of sources of coffee, ways of roasting beans and ways of preparing coffee are on offer; what type of coffee you drink and how you drink it supposedly signal your lifestyle and fashion sense. However, as a global commodity, coffee prices can be highly volatile. Small ripples in price in international markets may slightly influence the cost of a cup of coffee, but could make or break small-scale coffee producers in developing countries.[341]

The economics of coffee shops and museums have come full circle. Museums – the modern descendants of Salter's cabinet of curiosities and the guardians of our cultural collections – rely on profits from coffee shops. Whereas once coffee houses attracted customers with museums, museums may now attract customers with coffee houses.

Maize

Zea mays L.; Poaceae

CONQUERING THE WORLD

Christopher Columbus's rediscovery of the Americas in the late fifteenth century had profound effects on Europeans. It not only established a two-way trade of goods, people and ideas across the Atlantic Ocean, but influenced the natural world on both continents, shattered ecclesiastical certainties and the faith of intellectuals in the veracity of Greco-Roman authorities, and changed food cultures. As people began to break away from dogmatic beliefs that plants were God-given, they started to wonder about the origins of different plants. Such speculations were not just academic exercises for natural historians; they concerned how human cultures and civilizations developed, and ultimately provided raw material for the Darwinian intellectual revolution of the mid-nineteenth century.

Maize, one of a small group of usually annual Mexican and Central American grasses, was deeply rooted in Amerindian cultures. As the first European explorers of the Americas – the fortune-hunters, warriors, priests and administrators – island-hopped through the Caribbean and marched across Central and South America, they found a large-eared corn-like grain being consumed and used everywhere. In the Caribbean, the now-extinct Taino people called the plant *mahiz*. In Central America, the pantheons of the Mayans, Aztecs and Toltecs were populated with deities receiving regular offerings of maize. The Toltecs and Aztecs believed Quetzalcoatl, god of wisdom and knowledge, discovered maize and invented tortillas. The Aztecs even had separate male and female gods, Xochipilli and Chicomecoatl, for young maize plants, while the Mayans had Yum Kaax.

Although perhaps reluctant to admit it, European adventurers encountered civilizations at least as sophisticated as the ones they had left. All complex civilizations have a problem feeding themselves, smoothing the peaks and troughs of the agricultural cycle. The Aztecs solved the problem of feeding the large, densely housed, urban population of their imperial capital Tenochtitlan (on the site of modern Mexico City) by demanding tribute from satellite regions and adopting sophisticated agricultural systems. The *chinampas*, artificial islands of soil and compost created on the beds of shallow lakes, were one such system. Astute planting of *chinampas*, using combinations of maize, beans and pumpkins, meant soil quality and nutrient levels were retained for decades,[342] until they were destroyed with the fall of the empire in the early sixteenth century.

Modern maize cultivars usually produce one or two large yellow cobs per plant, but traditional Latin American maize cobs are tremendously diverse, both in the size and in the number of cobs per plant and in the kernel colours, which range from white through yellow and red to brown and black. When maize first arrived in the Old World it was unappreciated. The sixteenth-century English herbalist John Gerard stated:

> The bread which is made thereof is meanly white ... it is hard and drie as bisket is ... it is of hard digestion, and yeeldeth to the body little nourishment, it slowly descendeth and bindeth the belly ... the barbarous Indians which know no better, are constrained to make a virtue of necessitie, and think it good food ... a more convenient foode for swine than for men.[343]

However, Europeans gradually came to appreciate maize as a food. In 1620, as European crops failed around Massachusetts, the Pilgrim Fathers survived only because the local Iroquois taught them to grow maize. Maize became incorporated into Western civilization; one of the many ways American encounters changed Europe.

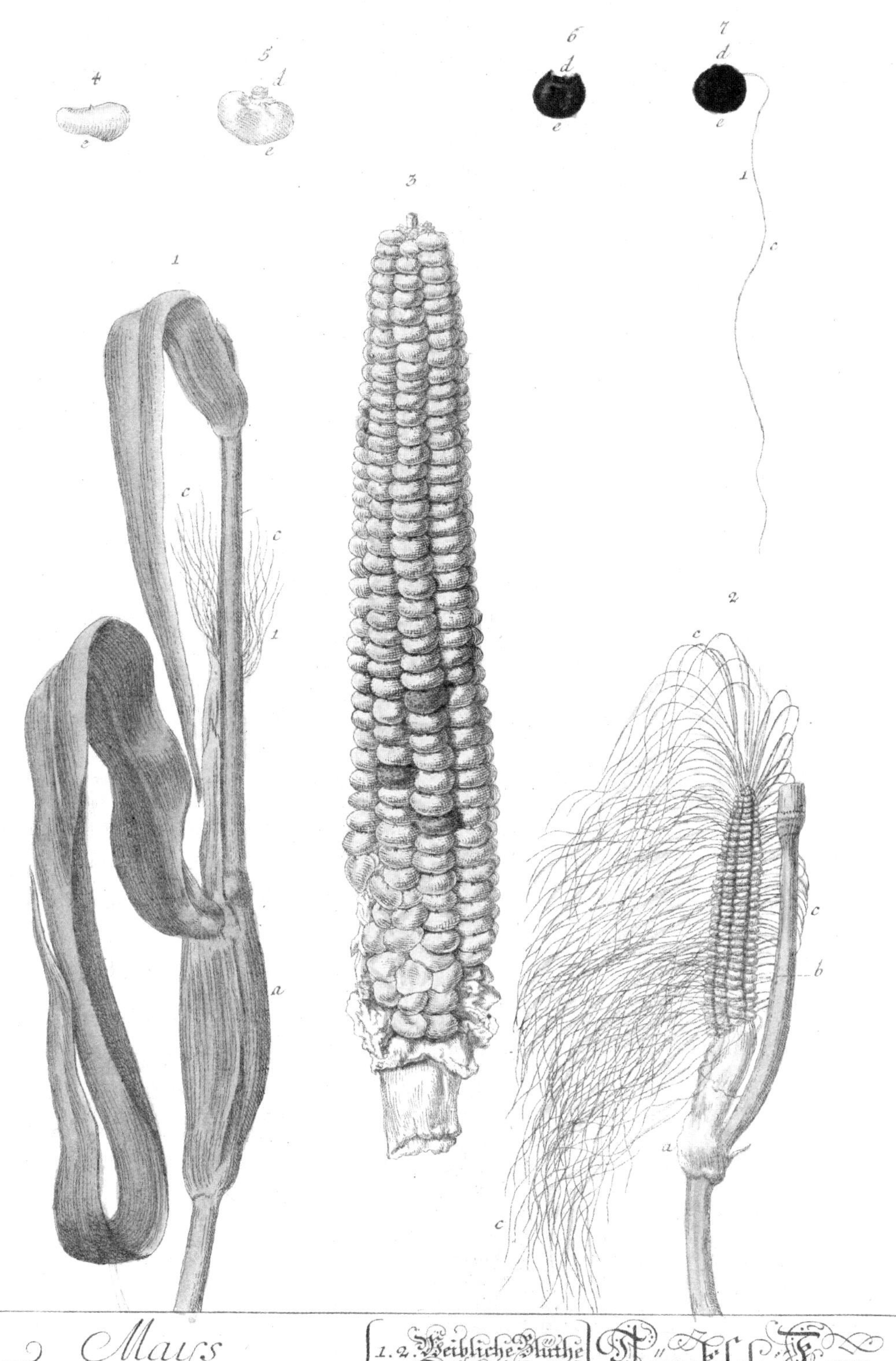

Mays
Frumentum turcicum.

1. 2. Weibliche Blüthe
3. Frücht
4–7. Saame

Türckisch Korn.

Maize has only ever been found associated with humans; it is unknown in the wild. It looks very different from all other American grasses, and consequently close wild relatives are not obvious. Two questions naturally arise. Did maize occur in the Old World before the discovery of the Americas? How did maize originate? Given the prominence of maize in the lives of indigenous Americans, and the antiquity of its cultivation, the question of maize origin has profound implications for the interpretation of the American archaeological record.

Evidence for the occurrence of maize in the Old World relies on interpreting names and descriptions in manuscripts and early printed sources.[344] The nineteenth-century Swiss botanist Alphonse de Candolle was uncompromising in his view that maize was American, yet his contemporary, the French agronomist Matthieu Bonafous, argued that maize occurred in the pre-Columbian Old World.[345] The Vinland Sagas, written down in the early thirteenth century, recount Leif Erikson's discovery of the New World some five centuries before Columbus. In 1885 one reference in the Sagas was equated to maize, throwing crop plant historians into confusion and reopening the possibility that maize was grown in the pre-Columbian Old World.[346] However, maize remains have never been recovered from Old World archaeological sites and there are no close Old World relatives of maize. References to seeds, interpreted as maize kernels, in pre-1492 sources are either forgeries or misinterpretations. European local names for maize, such as Barbary corn, Sicilian corn and Egyptian corn, emphasize where people believed maize came from, rather than where it originated. Overall, the picture is clear: maize is one of a handful of crops whose widespread Old World distribution occurred *after* 1492.

The moist, warm Mesoamerican climate makes the preservation of plant remains unlikely – but not impossible. A chronological sequence of thousands of cobs and cob fragments was discovered in the caves of the Tehuacán Valley in south-western Mexico. These caves contain layers of human refuse built up over thousands of years. By careful excavation, layers have been peeled away and the plant remnants retrieved, identified and dated.[347] Some of the oldest cobs are little more than the length of the first joint of your thumb, with eight rows of six to nine kernels. Significantly, each kernel is partially enclosed by a husk-like sheath.

The sole ancestor of maize is thought to be a Mexican wild grass, teosinte (literally 'grain of the gods'). Modification of just three genes transforms the appearance of teosinte.[348] Flowerless teosinte looks like branched maize, but when the kernels mature the differences between teosinte and maize are stark. Teosinte kernels are arranged in two-sided ears, with single kernels surrounded by hard shells, which break apart. In contrast, maize kernels are arranged in multiple-sided cobs of soft, paired kernels, which remain attached to the cob. Maize cannot disperse its seeds naturally, but teosinte seeds are specialized for dispersal through animals' guts.

Maize had taken about 9,000 years to cover the New World; following its introduction to Iberia, it took 400 years to cover the Old World. Today, millions of hectares of maize, much of it genetically modified, are grown worldwide, producing hundreds of millions of tonnes of kernels annually. Maize is a worldwide commodity, with traders speculating over maize futures, producing price spikes that influence food prices across the planet. Most corn is used for the production of animal feed and, controversially, biofuels. Golden-yellow maize cobs and kernels crushed and toasted for cornflakes, exploded for popcorn or pulverized for polenta are familiar foods, while purified maize starch and syrup are frequent food additives in almost all processed foods. Maize, under guiding human hands, has moved a long way from the caves of the Tehuacán Valley.

Ananas folio angusto serrato.

Pineapple

Ananas comosus (L.) Merr.; Bromeliaceae

HORTICULTURAL ONE-UPMANSHIP

One-upmanship, the desire to outdo one's neighbours for individualistic or nationalist reasons, is one of our common foibles. Garden shows, where people compete for having grown the biggest this or the heaviest that are popular pastimes, to which people devote vast amounts of energy. Few plants have inspired Europeans to show off more than the pineapple. In 1758 writer James Ralph summarized the pineapple obsession of British aristocrats: 'all must have their Fooleries as well as their Pinaries; and the only struggle seems to be, whose Fruit shall be the largest and most talk'd of.'[349]

By the middle of the eighteenth century the pineapple had been known to Europeans for a little over 250 years, since Christopher Columbus returned to Spain from his second voyage to the Caribbean. Only one of Columbus's pineapples was suitable for presentation to King Ferdinand and Queen Isabella but, according to an eyewitness, the Italian cleric and scholar Pietro Martyr d'Anghiera, 'the King prefers [pineapple] to all others'.[350] The pineapple had its royal imprimatur; it was special, suitable only for the very best in society. Royal, and divine, associations with the pineapple were drawn most tightly in Portuguese cleric António do Rosário's allegorical interpretation of Brazilian plants at the start of the eighteenth century ('Fruits of Brazil in a new, ascetic monarchy, consecrated to Our Lady of the Rosary').[351]

The pineapple is apparently native to southern Brazil and Paraguay, but domestication and widespread use by the peoples of pre-Columbian South America and the Caribbean have obscured its native distribution. In the mid-sixteenth century French traveller Jean de Léry encountered living pineapples in the area of present-day Rio de Janeiro, describing them by reference to familiar plants:

> leaves ... like the aloe's. It grows compacted like a great thistle; its fruit, related to our artichoke, is as big as a medium-sized melon, and shaped like a pinecone, but does not hang or bend to one side or the other.[352]

Botanically, pineapple is a compound fruit, made up of fused, fleshy berries produced by individual flowers; the armoured outside comprises the remnants of flower parts and bracts. Pineapples are propagated vegetatively from the crown of leaves surmounting the fruit or from shoots at the fruit or plant base.

Early pineapple descriptions focused on the fruit's beauty and sweetness. Léry was not unusual in the superlatives he applied to the 'iridescent yellow' fruit with 'a fragrance of raspberry' and a taste that 'melts in your mouth, and is naturally so sweet that we have no jams that surpass them; I think it is the finest fruit in America'.[353]

Pineapple cultivation spread rapidly through the Old World as the hegemony of the Portuguese, under the terms of the Treaty of Tordesillas (see NUTMEG, page 137), extended Portugal's reach into tropical Africa and Asia. By the mid-1600s the pineapple was encountered frequently enough in China for the Polish missionary and explorer Michael Boym to mistake it for a Chinese native.[354]

In 1657 colonist Richard Ligon wrote the first detailed, eyewitness description of pineapple in English, and pineapples were first seen in England when some were presented to Oliver Cromwell.[355] However, absence of first-hand knowledge had not prevented English authors illustrating and describing the wonders of the pineapple to their readers.[356] When diarist John Evelyn tasted pineapple, at King Charles II's table on 10 August 1668, he was disappointed; the reality of its flavour did not live up to the hype of the literary descriptions.[357]

The British and Dutch were competing for dominance in the commerce and horticultural spheres during the seventeenth century. The Dutch had supremacy in both, so for the British to be able to cultivate a home-grown pineapple would have been a coup. In about 1670 Charles II was painted receiving a pineapple from his gardener, John Rose. It reinforced royal associations but, despite the implications, it is very unlikely pineapples were being raised in seventeenth-century England. However, in about 1714 the Dutch gardener Henry Telende, using the vast wealth of businessman Matthew Decker, did succeed in growing English pineapples. Telende's 'trick' was to use the heat generated by the fermentation of dung and tanners' bark. When Cambridge professor of botany Richard Bradley published in detail Telende's methods,the Georgians now had an instruction manual and could proceed apace with the 'limitless' patronage of the landed gentry.[358] Even the universities of Oxford and Cambridge were not immune to the pineapple craze; pineapples fruited in the Oxford Physic Garden in 1744 and at Trinity College, Cambridge, four years later. By the end of the century gardener William Speechly published *A Treatise on the Culture of the Pineapple* (1779), the definitive eighteenth-century account on pineapple cultivation. Literature on pineapple husbandry blossomed into the nineteenth century as Victorians took over the fascination with pineapple cultivation, an obsession shared with gardeners in North America and continental Europe.

Pineries became the playthings of the vastly wealthy, and the ability to deliver ripe, home-grown pineapples to banqueting tables became a mark of social and intellectual distinction, along with having a fine garden and well-stocked library. The pinery, and its contents, became symbolic of the rights of the landed gentry to rule and displayed power over nature; even in a cold, dark land such as Britain, tropical pineapples could be made to produce fruit.

The less wealthy upper classes had their glasshouses where they could compete at growing other plants, and became enamoured by all sorts of pineapple paraphernalia and tat, such as teapots, snuffboxes and stone garden adornments. However, few garden ornaments matched for sheer grandiosity the pineapple-inspired folly that still stands in the grounds of Dunmore Park, Scotland.

By the end of the nineteenth century, technology and transport were making pineapples less socially exclusive, and the complexity and effort of home-grown pineapple production – the maintenance of gardens and pineries required large numbers of skilled gardeners and a great deal of money – were becoming an anachronism.[359] This was perhaps not fully appreciated until after the First World War, when the British aristocracy had to come to terms with the slaughter of young gardeners on European battlefields and the chill of economic reality.[360]

The pineapple has lost none of its exoticism but advances in transportation and preservation technologies have made it commonplace, whether fresh, canned or juiced. Canning in particular democratized pineapple consumption but changed the flavour and the appearance of the fruit; pineapples needed to be bred that fitted conveniently into tins. Our desires to impress have not changed; they have been transferred to other things – some botanical, many not.

Smooth meadow grass

Poa pratensis L.; Poaceae

PERFECT PASTURE PLANT

We may not graze on grass ourselves, but we do rely heavily on grazing animals for meat and dairy products. Before the invention of the internal combustion engine, animals, largely fed on grass, were the power for transport and heavy agricultural labour; prized products from the East moved into Europe on the backs of camels and mules along the Silk Roads; forests were transformed to arable land using the muscles of oxen and horses. Even today, more than 3.7 trillion asses, buffaloes, camels, cattle, goats, horses, mules and sheep provide us with food, transport and fertilizer.[361] No matter how we use them, livestock must be fed. Consequently, there are intimate bonds linking the lives of plants (especially grasses), livestock and humans: no plants, no livestock; no livestock, no manure; no manure, no plants.

The way to make a living from grazing animals is to find or grow the right fodder (usually a grass) in the right place, and manage the land in the right way. As a North American prairie farmer asserted with characteristic mid-nineteenth-century self-reliance: 'whoever has limestone land has blue grass; whoever has blue grass has the basis of all agricultural prosperity; and that man, if he have not the finest horses, cattle, and sheep, has no one to blame but himself.'[362] However, the wonderful bluegrass was not a North American native; it had arrived from Europe under a different name.

Common plant names enjoy neither uniformity nor international recognition; instead they reflect local cultures, traditions, beliefs and prejudices. In the official British list of plant common names bluegrass is known as smooth meadow grass. In other parts of Europe this grass is called *pâturin des prés* (France), *capim-do-campo* (Portugal), *grama de prados* (Spain) or *Wiesen-Rispengras* (Germany). All mean, more or less, 'meadow grass'. Scientific names, on the other hand, aim to be unambiguously understood across language barriers; they promote international communication. This is only true of how we apply scientific Latin names today, which is thanks to the Swedish naturalist Carolus Linnaeus. Before Linnaeus, this grass sported a multitude of complex Latin names, such as *Gramen pratense paniculatum majus, latiore folio* (the large-panicled meadow grass with broad leaves). Linnaeus's innovation was the consistent application of a simple binomial naming system. In 1753 he formally named this grass *Poa pratensis*. The first name, *Poa* (from the Greek meaning 'grass'), is the genus name and refers to a group of more than 500 species. It is the second word, *pratensis* (meaning 'of the meadow'), that uniquely identifies a particular species. But even within the species, smooth meadow grass is highly variable, with many different genetic variants. Some of these variants accidentally crossed the Atlantic with early European settlers, who valued it as a fodder grass.[363] It competed well with native American grasses and, just like the people, adapted and flourished. Eventually, as the settlers moved west and smooth meadow grass followed in their wake, vast areas of prairie were colonized without people even noticing. Those strains, selected in American soils first by trial and error and then by systematic experiments, transformed the prairies. And so the legendary Kentucky bluegrass came to be.

The unidentified landowner quoted above was not the first to wax lyrical about bluegrass. In the 1870s Senator John Ingalls's prose became purple with local pride, as he spoke of the importance of this special Kentucky forage grass. In the meantime, enthusiastic European landowners, such as the sixth Duke of Bedford, whose family seat is Woburn Abbey, were also striving to improve grazing grasses and

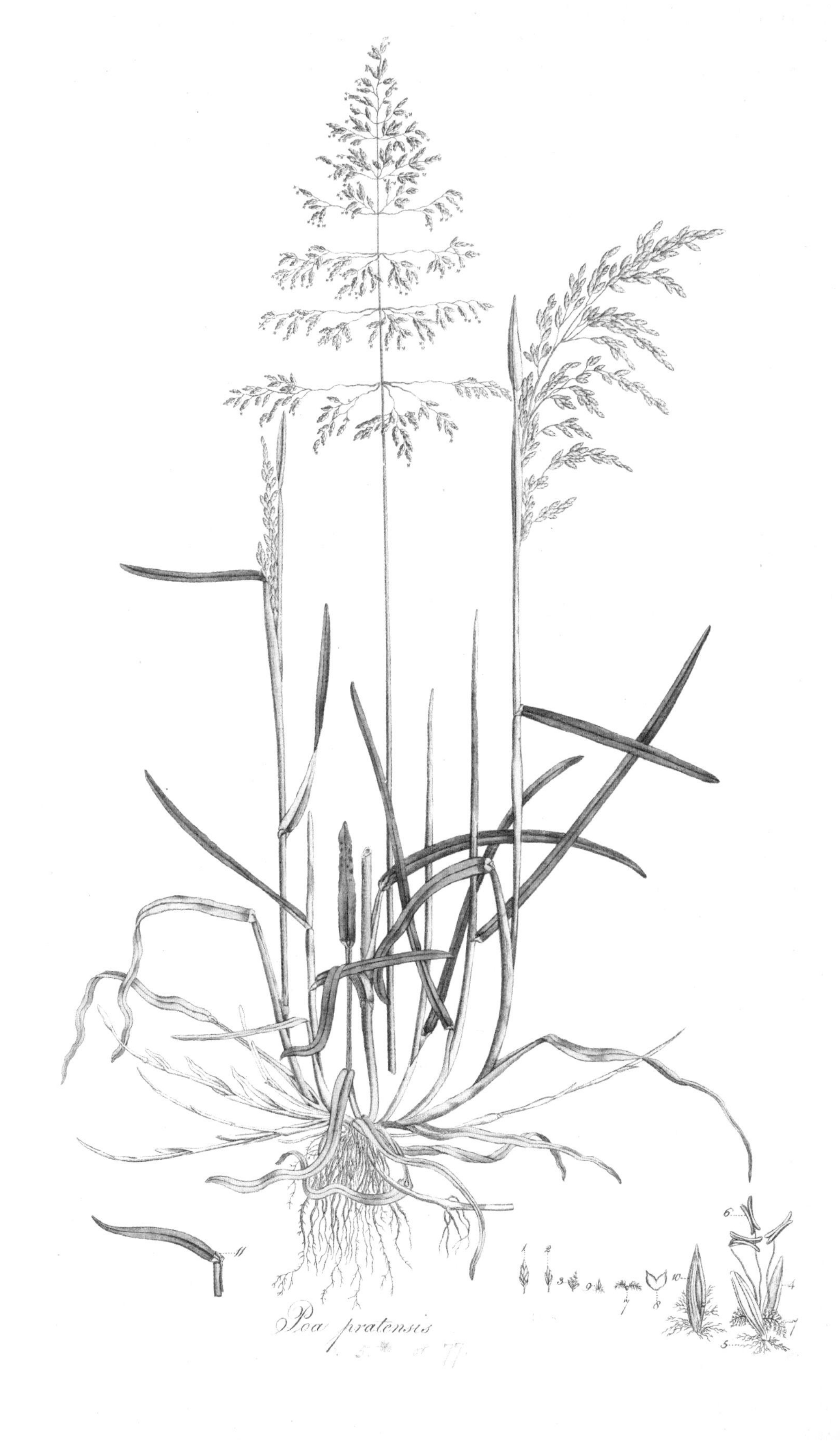
Poa pratensis

started to publish their results. George Sinclair, in *Hortus Gramineus Woburnensis* (1816), considered smooth meadow grass overrated as a fodder compared to other grasses. However, true to the adage that grass is always greener on the other side, American Kentucky bluegrass was eventually imported back to Europe to improve smooth meadow grass as forage and turf.

The fodder quality of grasses is determined by nutrient and calorie content. In Europe, when temperatures fall below 5°C, grasses hardly grow at all, while in warm, wet periods they grow vigorously, although as grasses get older overall fodder quality decreases. Consequently, farmers and graziers must capitalize on the production of nutrient-rich young leaves. They also have to take into account the different ways animals graze. Cattle, for example, rip up tufts of grass, producing an uneven, tussocky sward, while sheep nibble at grasses, producing a short sward.

To feed livestock over the lean months, grasses must be stored. Forage grasses can be stored either dry as hay or wet as silage. In both cases, they must be harvested at their maximum nutritional value. As farmers have long known, to produce the best hay the crop must be dried quickly and thoroughly before storage. Given the vagaries of late summer weather in many areas, 'pickling' grass as silage has become the preferred method of preservation in many Western European agricultural systems.

The demands of livestock have meant virtually all grasslands in Western Europe are products of human activities over the last five millennia. Rough grazing, of the uplands or marginal lowlands, tends to be rich in grasses adapted to growing in low-nutrient conditions, while permanent pasture, the mainstay of much livestock farming, has been created by centuries of 'improvement' of lowland rough grazing through drainage and applications of fertilizer.

Nitrogen is critical for plant growth and must be added to most soils to increase plant productivity. Humans have tried all manner of ingenious, often macabre, ways of adding it to the soil. By the mid-eighteenth century, large-scale transformation of rough grassland to permanent pasture was under way in Western Europe, land was drained and – knowledge of the power of muck to improve plant growth having been handed down from generation to generation – fertilized with all manner of plant and animal waste products, including flesh and bones, bile, mucus, urine and ordure; organic matter for fertilizer was at a premium. However, this changed completely in the twentieth century with the invention of the Haber–Bosch process. This method takes nitrogen directly from the atmosphere and combines it with hydrogen to produce ammonia, the raw material for industrial fertilizer manufacture. Nitrogen became available in unlimited supply, and the productivity of grasslands increased dramatically. Grassland species evolved in conditions of low fertility. Once grasslands were covered with fertilizer many species could no longer compete with the few grasses that could make use of highly fertile conditions. As agricultural production intensified, rough grassland, including flower-rich hay meadows, declined across the Western world.[364]

The twentieth century also saw grasslands reduced from a different cause. In the interwar years, as draught animals gave way to petrol and diesel engines, so land was liberated from the production of fodder. Economic realities meant pastures were ploughed and hedgerows grubbed up to make way for fields of wheat. With the rise in public awareness of species conservation, grasslands, especially hay meadows, have become a focus for the loss of familiar species (see CORNCOCKLE, page 266).

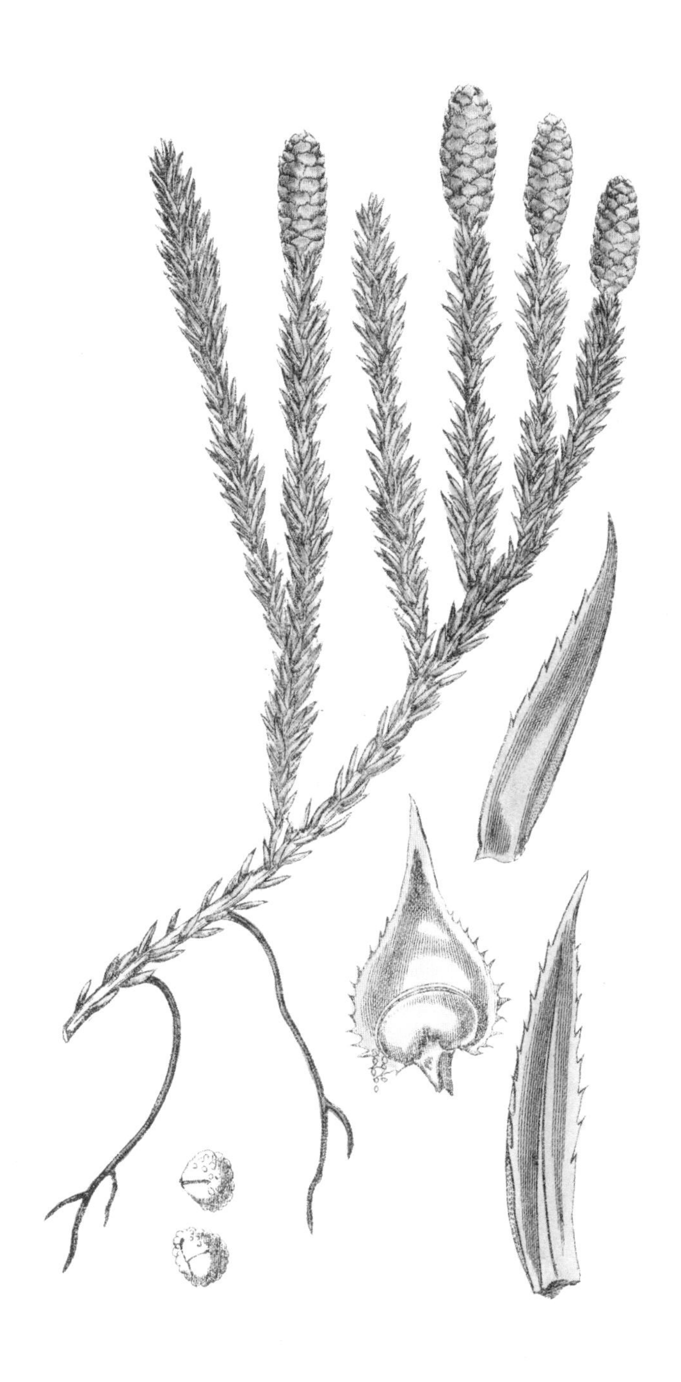

Lycopodium Annotinum.

Lycopods

Lycophyta

RELEASING FOSSILIZED STARLIGHT

Lycopods? Why choose clubmosses and their cousins? Well, they have done rather a lot for us: electricity and energy for transport, raw materials for heavy industries, the chemicals for fertilizer, pesticide and medicine manufacture, and more besides.

During the Carboniferous period, 300 to 350 million years ago, this planet was very different from that which we see today. Large parts of North America and Europe were located near the equator, and covered by vast forested wetlands. Modern humans were hundreds of millions of years from evolving, yet it was within these shallow, tropical and subtropical swamps that the coalfields that were to fuel the industrial development of modern Western civilization were formed.

The plants of these tropical wetland forests were not the flowering plants that dominate tropical forests today, but relatives of horsetails and ferns, together with early gymnosperms (non-flowering plants) and extinct groups such as seed ferns. However, the most significant group was the lycopods, which today are broadly recognized as spike mosses, clubmosses and quillworts. Lycopods, or at least their prehistoric cousins, built the modern world.

When these prehistoric forested swamps were naturally flooded, the anaerobic conditions slowed decay, and the partially rotted plant material gradually became peat. Layers of sediment washed over the peat, crushing it, slowly raising its temperature and eventually carbonizing it. Over geological time, plant remains in these Carboniferous peats were gradually converted to coal: lignite and finally anthracite. Billions of tonnes of coal and associated gases were trapped, some close to the Earth's surface, others deep underground.

The fossilized remains of Carboniferous lycopods are frequently found in coal measures. They were giants, some reaching the height of ten-storey buildings, with trunk diameters about the width of a big car. Their living descendants, approximately 2,000 species of them, are much smaller and found in areas as diverse as temperate and tropical forests and savannas, tundra and montane grasslands. The largest genus is *Lycopodium*. These are short, branched plants, covered with needle- or scale-like leaves, with cones of male and female spores arranged in clusters on the branch tips. The cone clusters can appear like small paws, fancifully imagined to resemble those of wolves, hence the scientific name.

In pre-industrial times most energy needs were supplied by water, wood, people or animals. However, coal was not unknown. The Ancient Greek philosopher Theophrastus described stones that were dug from the earth and could be burned to work metal. Bronze Age Britons used coal, while Romans exploited most English coalfields and used coal across Europe to smelt metal ores. In the thirteenth century, when the Venetian merchant Marco Polo travelled to China he described how the Chinese used black stones that burnt like charcoal. Until the mid-eighteenth century the majority of coal came from surface deposits. With the British Industrial Revolution and the development of deep mining technologies, coal became a major energy source; our addiction to fossil fuels had begun. The industrial clamour for coal, first in Britain and then throughout the West, triggered dramatic political, social, environmental and economic changes, as markets for mass-produced goods were created, people left villages to crowd into cities, and social units were fractured. The ripples of these fundamental changes have continued across the globe to the present day.

In addition to being an energy source, coal is an industrial raw material. Heating coal in an oxygen-free atmosphere converts it to coke, which may be used in blast furnaces to extract iron from iron ore. Gases released by heating coal produce flammable, highly toxic coal gas – an important fuel before the exploitation of natural gas. Coal can also be used to manufacture building blocks for industrial chemistry, such as methanol and ammonia, which came to the fore in the late nineteenth century.

Coal consumption comes with social and environmental costs. Coal mining is dangerous, killing hundreds of people a year and maiming thousands, often with crippling lung diseases. Environmentally, coal mines and coal-based industries pollute water supplies and produce thousands of tonnes of toxic waste. Where large amounts of coal are burnt, the result is smoke rich in soot and sulphurous gases. Combine this with the fogs typical of northern temperate winters, and you get a dangerous sulphuric, acid-rich cocktail: smog. Smogs slowly dissolve the architectural monuments of civilizations and, not so slowly, kill people. London smogs, frequent during the nineteenth and twentieth centuries, were so thick they became known as 'pea-soupers'. The most famous of these was the 1952 Great Smog, which killed some 8,000 people over a matter of months.[365] With the introduction of the 1956 Clean Air Act, smogs associated with coal fires were consigned to history, and hundreds of years of architectural heritage, dissolving in these acidic mists, could be cleaned and conserved for future generations to enjoy.

Extant lycopods do not have many modern uses, but those they do are surprisingly varied. The fine, light, yellow-brown spores produced by the cones repel water extremely well and are highly flammable. Consequently, they have been used in flash powders and fireworks as entertainment, in fingerprint powders to apprehend criminals, in ice cream as stabilizers and as condom lubricants to prevent latex allergies. In 1938 lycopodium spores even played a role in the development of early photocopiers, paving the way for an information revolution.[366] This short list is perhaps an interesting reflection of preoccupations in Western civilizations.

Products from prehistoric lycopods continue to create profound dilemmas for all cultures. The remains of fossilized lycopods are the planet's most important source of energy for electricity generation. There are billions of tonnes of coal reserves in the Earth's crust, which, if burned, would generate electricity for thousands of years. This presents a dilemma. In coal deposits, carbon dioxide, one of the main greenhouse gases, is locked away in a form that cannot affect the atmosphere. However, once coal is burnt, millions of tonnes of carbon dioxide and other toxic gases are released, affecting the planet's temperature and atmospheric quality. These issues are at the heart of current arguments over the use of scientific evidence to inform energy policies and the abilities of politicians to produce long-term, evidence-based environmental policies. Can we afford to run our lives on the capital of prehistoric photosynthesis? Can we afford not to?

Cotton

Gossypium species; Malvaceae

CLOTHING THE GLOBE

In 1860 the American South and Britain were bound together by the wealth created by one plant: cotton. For forty years the agrarian slave states of the American South had grown, harvested and supplied raw cotton to the spinners and weavers in the cotton mills of industrial Lancashire, feeding the world's gluttonous appetite for cotton products. Bare numbers give only a glimpse of the importance and intimacy of the Anglo-American relationship with King Cotton on the eve of the American Civil War. More than 3,000 British cotton mills, employing some 650,000 people, processed approximately 450,000 tonnes of raw cotton, of which about 80 per cent was imported from the American South and worth to these states nearly $120 million dollars.[367] At the beginning of the nineteenth century America supplied about one-tenth of the world's cotton; by 1860 this had risen to two-thirds of the total. Global production of cotton had risen nearly fivefold over the same period. By 1860, India, traditionally the source of the world's cotton, barely supplied 20 per cent of the need.

Cotton plants are a distinctive sight when the mature fruits, called bolls, each split open to reveal a gossamer-fine white mass of fibres the size of a golf ball. These fluffy cottonwool spheres are thought to be at least partially responsible for the myth of the barometz, or vegetable lamb. This late medieval chimerical creation, of a sheep attached to a stalk, was fostered by the gullible well into the early modern period.[368] Even without such distractions, discerning real cotton in classical Western literature is mired in a miasma of misapplied names.[369] The Greek philosopher Theophrastus knew cotton from India and

Bahrain,[370] but despite the modern cachet of Egyptian cotton there is little evidence of cotton in Egypt before the Islamic period; Ancient Egyptian 'cotton' was made from flax.[371]

The genus *Gossypium* contains about fifty perennial species, found across the globe's tropical and subtropical regions.[372] Cotton has been separately used and domesticated in both the Old World and the New World for at least 5,000 years. Between 1 and 2 million years ago cotton plants emerged in the Americas that were hybrids of a Peruvian cotton species and a close relative of modern-day Old World *Gossypium herbaceum*.[373] These New World hybrids diverged into numerous species, from which today's commercial cottons were domesticated by multiple indigenous New World cultures: *G. hirsutum* in Central and South America and *G. barbadense* in the Caribbean. Before 1800 the Old World species, *G. arboreum* and *G. herbaceum*, were the most important commercial sources of cotton. Today's Egyptian cotton is a nineteenth-century selection of a New World species.

The fibres of cotton, known as staple, are hollow tubes, the remains of the cellulose-rich walls surrounding the giant hair cells covering the surface of the top-shaped boll. The wall of the fibre is layered. The outer layer is a smooth cuticle, inside which are concentric cellulose layers that give the fibre a natural twist; together these features produce a smooth, readily spun fibre. Approximately 200 million fibres are needed to make 1 kg of raw cotton. The conditions under which particular cotton cultivars grow influence the structure of the fibres and give cotton its mechanical properties.[374] The seed fibres of cotton's wild ancestors are short, producing a felt-like covering to the seed surface. During domestication, humans selected long staple mutants, with a preference for brilliant white fibres over the dirty white fibres of the wild types. Long staple cotton is more easily spun than short staple cotton, so Old World 'commercial' cottons, with shorter staples than their American cousins, became undesirable as cotton processing was industrialized – in cultivated American cotton, individual fibre

a. Gnaphalium seu Ely-
chrysum angustifolium luteum. b. Gnaphalium
montanum, seu Pes-cati, flore candido. c. Gnaphalium montanum flore rubro.
d. Gnaphalium montanum flore rotundiore.
e. Gossypium arboreum flore atro purpureo, Cotton, Baumwolle
f. Gossypium frutescens annuum.

cells can measure 3 cm. Other desirable cotton plant features, modified through domestication, include small seed size, easily detached fibres and an annual growth habit.[375]

An anonymous nineteenth-century English army officer expressed a common view that cotton was 'raised more perhaps by the bounty of Providence than the labour of Mankind'.[376] Such a comment might be excused of an ignorant inhabitant of a Victorian drawing room; it ill behoved someone familiar with British India, with knowledge of the reality of the hard labour needed to produce cotton fabric. Inhabitants of fashionable salons from Boston to Paris gave little thought to conditions under which slaves were forced to work in the cotton plantations of the American South to produce raw cotton, or the circumstances under which workers in northern British mill towns laboured.

Mechanization was a key innovation to producing affordable cotton fabrics – before industrialization, cotton-based textiles were the most expensive available. In eighteenth-century England, Lancashire became the centre of the cotton industry as a steady stream of inventions boosted cotton manufacture: Lewis Paul and John Wyatt's roller spinning machine, James Hargreaves's spinning jenny, Richard Arkwright's spinning frame and Samuel Crompton's spinning mule. In 1793 the American Eli Whitney patented a cotton gin for rapidly separating fibres and seeds in the cotton boll. The landscape of England changed as the Industrial Revolution gathered pace, while the American South expanded as vast areas of land west of the Appalachians were transformed under cotton culture.

The cotton fields made the American South economically viable, and to preserve their slave-based economy the Confederate States seceded, which led to the American Civil War in 1861. While free traders, abolitionists and apologists for slavery argued among themselves, the leaders of the American South took a calculated gamble to withhold the cotton crop, in the expectation that the British government would support their cause.[377] However, British cotton stocks were at an all-time high and there was a general depression in the textile industry because of contracting markets; belligerent mill owners objected to curbs on free trade and sided with the Unionists. The result was the Lancashire cotton famine.[378]

As the cotton famine took hold and unemployment in Lancashire increased, the British tried to establish alternative sources of cotton supply from her colonies, although at the expense of staple quality. Following the end of the Civil War, American cotton supplies rapidly resumed, employing a system of black and white sharecroppers, and the Lancashire mills started to churn out cotton goods once more. However, the Indian and Egyptian populations were to suffer dreadfully towards the end of the century through the iniquities of the cotton trade and British attempts to diversify their cotton suppliers.[379]

The main use of cotton is for soft, 'breathable', absorbent fabrics and yarns, but other uses include production of high-quality acid-free paper and the manufacture of nitrocellulose, an explosive used to make early photographic film, theatrical flash paper and sticky surfaces for catching DNA. Even cotton seed does not go to waste: an edible vegetable oil can be extracted and the residue used for cattle feed. Cotton plants also contain the toxic pigment gossypol, which appears to affect human reproduction.[380]

In 2022 an area about half the size of France produced approximately 27 million tonnes of cotton fibre; together China and India accounted for more than a third of global cotton production. Cotton today comes with similar economic, political, environmental and social concerns as those of previous generations. Economically and politically, cotton markets are volatile, subject to speculation, hoarding and the vagaries of the weather and fashion. Environmentally, cotton production needs large amounts of land, water, fertilizers, pesticides and herbicides to produce viable crops, although the creation of insect-resistant and herbicide-tolerant genetically modified cotton is argued to have environmental benefits.[381] Socially, cotton production has been strongly associated with modern slavery issues in developing countries, as consumers demand cheap clothing. Cotton is once more associated with exploitation.

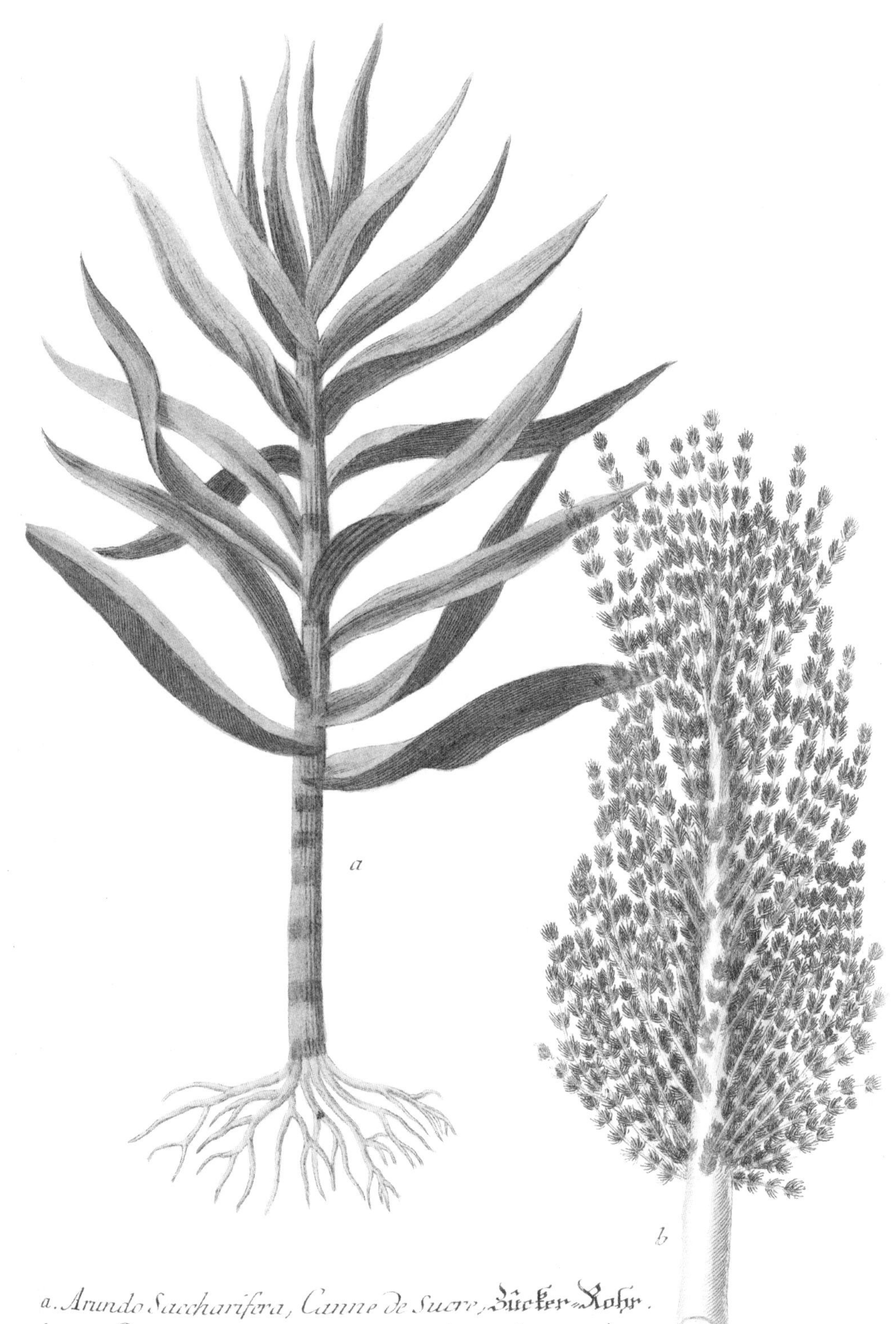

a. *Arundo Saccharifera, Canne de Sucre,* Zücker-Rohr.
b. *Coma Arundinis sacchariferæ, fleure à Canne de Sucre,* Aehre von Zücker-Ried.

Sugar cane

Saccharum officinale L.; Poaceae

FORTUNE MAKER AND ENSLAVER

In the course of 3,000 years, sugar cane has been transformed from a Papua New Guinean 'kitchen garden' plant into a global agro-industrial crop. In the process, the Americas, Africa and Europe were changed, cultures exterminated and populations enslaved. In 1711 Italian Jesuit Giovanni Antônio Andreoni was plain: 'slaves are the hands and feet of the sugar refinery owner; without them in Brazil it is impossible to make, preserve and increase the plantation, nor have a refinery.'[382] Before the seventeenth century Europeans had little experience of concentrated sweetness in the form of the carbohydrate sucrose; they used honey. In medieval Europe, sugar was rare and consequently very expensive. By the sixteenth century the cost of sugar was merely exorbitant and being used by the very wealthy to garner favour with the extremely powerful. Eventually, sugar became what it is today: a commonplace preservative, foodstuff and sweetener.

About half the sugar consumed today is extracted from a perennial grass, sugar cane, the 'sublimest effort of heat and light'.[383] Carolus Linnaeus gave the name *Saccharum* to this genus of tropical and subtropical grasses because of sugar cane's sweet stems. Sugar cane is known only from cultivation, and probably originated in Papua New Guinea. As people migrated across Polynesia and into Southern and South East Asia the cane travelled with them. By the sixth century sugar cane was being grown in Persia; by the middle of the eighth century it was flourishing in Egypt, and by the tenth century it was an important Middle Eastern crop. Arab expansion took sugar cane throughout the Mediterranean as far as the Iberian Peninsula, and by the fifteenth century the Spanish and Portuguese had introduced it to their island possessions in the Atlantic (nowadays collectively labelled Macaronesia), from where it was transferred to the Americas.

Extracting and concentrating sucrose from cane is a complex, skilled process. The juice must be squeezed from the cane stem and then boiled and reduced. When the concentrated syrup cools, the sucrose crystallizes and the molasses drains away. A field the size of a football pitch will produce about 50 tonnes of sugar cane, equivalent to some 6.5 tonnes of refined sugar; the average annual sugar consumption for 270 people in the United Kingdom.

Sugar cane, which John Gerard called a 'pleasant and profitable reed',[384] makes great demands on the growers and the environment. Land must be deforested, large amounts of water made available and high levels of fertilizer applied to make cane fields productive. Planting and fertilizing fields, harvesting cane and extracting sugar are labour-intensive processes. Furthermore, sugar cane grows best in areas with high temperatures and rainfall and rich, deep soils, the areas where manual labour is most difficult – sugar cane history is one of human exploitation and environmental destruction, in which slavery played a major role.

Christopher Columbus introduced sugar cane to the New World from the Canary Islands, on his second voyage in 1493. He supervised the cane planting on the island of Hispaniola (present-day Dominican Republic and Haiti) and was amazed at the rapidity with which it grew. Hispaniola was populated by some 2 million indigenous agriculturalists called the Taino, but agriculture was not a major concern for the new settlers; they wanted the immediate reward of gold. By 1542 Bartolomé de las Casas, the chronicler of Spanish abuses in the West Indies, recorded only 200 surviving Taino. Within two decades the Taino were extinct, wiped out by the deadly combination of forced labour, religious dogma, bigotry and imported diseases, such as smallpox, measles, cholera and influenza. Today, all that remains of Taino culture are isolated objects in museums and the descriptions in Las Casas's accounts. The Taino were among the first victims of sugar cane cultivation in the Americas. They were not the last.

Once the worth of sugar cane was recognized, Iberian and British landowners across the Americas brought in African slaves to work the fields, and the Transatlantic slave trade became established. Without a slave labour force, economic sugar production would have been impossible. At the end of the eighteenth century the English plantation doctor and sugar enthusiast Benjamin Moseley succinctly summarized the position:

> If Jamaica, and the other English sugar islands, were to share the fate of *St. Domingue* [Haiti], by the horrors of war, a distress would arise, not only in England, but in Europe, not confined to the present generation, but that would descent to the child unborn. Of such important has the agriculture of half a million of Africans, become to Europeans.[385]

Voltaire's horrified hero in *Candide* (1759) reports meeting a double-amputee slave on the side of a Surinamese road, who tells him: 'when we work in the sugar refineries and catch our finger in the mill, they cut off the hand: when we try to run away, they cut off a leg: both things have happened to me. It is at this price that you eat sugar in Europe.'[386]

By the end of the eighteenth century the morality of the slave trade was being questioned, but the profits that rolled in on the backs of slaves and indentured labour quelled consciences. Only during the nineteenth century were laws framed by European empires that outlawed slavery. By this time Europe was addicted to sugar, whatever the cost.

As well as a sweet blessing, sugar was regarded as a medicine. Although some associated sugar consumption with the 'pissing evil' (diabetes), the effects of sugar on teeth were dismissed as little more than 'a prudent old woman's bug-bear, to frighten children'.[387] Sugar was not only killing the producers; it was killing its consumers, albeit sweetly and silently. Today, sugar's killing power is subtly manifest through obesity, type 2 diabetes and heart disease, not to mention bad teeth. As concern over sugar has risen, so substitutes have become fashionable, including artificial sweeteners such as aspartame, or alternative natural sweeteners, such as leaves of the South American sweetleaf daisy, *Stevia rebaudiana*.

Sugar made Britain a fortune. West Indian sugar indirectly funded the development of cities such as Liverpool and Bristol, the establishment of the Royal Botanic Gardens, Kew, and buildings such as the Tate Gallery in London. West Indian cane plantations also provided convenient places for younger sons of the landed gentry, or members of the emerging middle class, to make their fortune. Sugar made beverages such as tea, coffee and chocolate palatable, leading to the rise of the tea and coffee shops as meeting places for political, artistic and scientific debate (see COFFEE, page 185). During the nineteenth century the cultural stereotype of Britain as a nation of sweet-tea drinkers was born; indeed, bread and sweet tea sometimes appeared to be staples for factory workers during the Industrial Revolution.

The supremacy of sugar cane as a sugar source was challenged as Napoleon successfully promoted sugar beet in the early nineteenth century (see BEETS, page 28). Today, both sugar cane and sugar beet face competition from the corn syrup extracted from grasses such as sorghum and maize.

Coconut

Cocos nucifera L.; Arecaceae

GLOBAL TRAVELLER

Victorian coconut-matting entrepreneur William Purdie Treloar prefaced his paean to coconut utility, *The Prince of Palms* (1884), with an unattributed epigraph: 'clothing, meat, trencher, drink, and can, / boat, cable, sail, mast, needle, all in one'.[388]

For millennia the coconut has been central to the lives of Polynesian and Asian peoples; the fruits being portable food and fluid, and the stems and leaves being shelter, fuel and canoes.[389] In Europe, on the other hand, coconuts have always been exotic and unusual, sometimes rare. Marco Polo apparently saw coconuts in South Asia in the late thirteenth century,[390] and among the marvellous, mid-fourteenth-century musings of the quixotic 'traveller' Sir John Mandeville there is mention of 'great Notes of Ynde' (great Nuts of India).[391] Uncharacteristically, the English herbalist John Gerard underestimated the utility of the Indian Nut tree, highlighting polished coconut shells, traced with silver, as novelty items for wealthy Europeans.[392] During Victorian and Edwardian Britain, coconut shies, where coconuts were both targets and prizes, became commonplace fairground entertainment.[393] Today, images of palm-fringed tropical beaches are clichés to sell holidays, chocolate bars, fizzy drinks and even romance.

Typically, we envisage coconuts as brown cannonballs that take ingenuity to open for the reward of their sweet white flesh. But we see only part of the fruit and none of the plant from which they come. The coconut palm has a smooth, slender, curved, grey trunk, up to 30 metres tall, surmounted by a rosette of leaves, each of which may be up to 6 metres long. Immature coconut flowers are tightly clustered together in a hard sheath among the leaves at the top of the trunk. The flower stems may be tapped for their sap to produce alcoholic palm wine (toddy), which in turn is distilled to produce arrack or dried to make an unrefined sugar, jaggery.

Coconut palms produce as many as seventy head-sized fruits per year, with fruits weighing more than 1 kg each. The wall of the fruit has three layers: a waterproof outer layer, a fibrous middle layer and a hard inner layer. The thick fibrous middle layer produces coconut fibre, coir, which is used for making rope, matting and bristles, and even serves as a peat substitute. The woody innermost layer (shell) surrounds the seed and is the familiar coconut of grocery shelves. The shell, with its three prominent 'eyes', gives coconut its scientific name, *Cocos*, meaning 'mask' or 'monkey face' in Portuguese and Spanish.

Inside the shell are the nutrients (endosperm) needed by the developing seed. Initially, the endosperm is a sweetish liquid, coconut water, which is enjoyed as a drink, but also used as a source of plant growth hormones and even, as it is sterile, as emergency blood transfusions. As the fruit matures, the coconut water gradually solidifies to form the brilliant white, fat-rich, edible flesh or meat. Dried coconut flesh, copra, is made into coconut fat and coconut milk, which are widely used in cooking, cosmetics and soap.

Their biology would appear to make coconuts the great maritime voyagers and coastal colonizers of the plant world. The large, energy-rich fruits are buoyant and salt-tolerant, but cannot remain viable indefinitely; studies suggest the limit is about 110 days.[394] Literally cast onto desert island shores, with little more than sand to grow in and exposed to the full glare of the tropical sun, coconut seeds must germinate and root. The air pocket in the seed, created as the endosperm solidifies, and the fibrous fruit wall that provided buoyancy *en voyage*, protect the embryo and provide a moisture-retentive rooting medium to give the coconut seedling the start it needs to establish itself quickly.

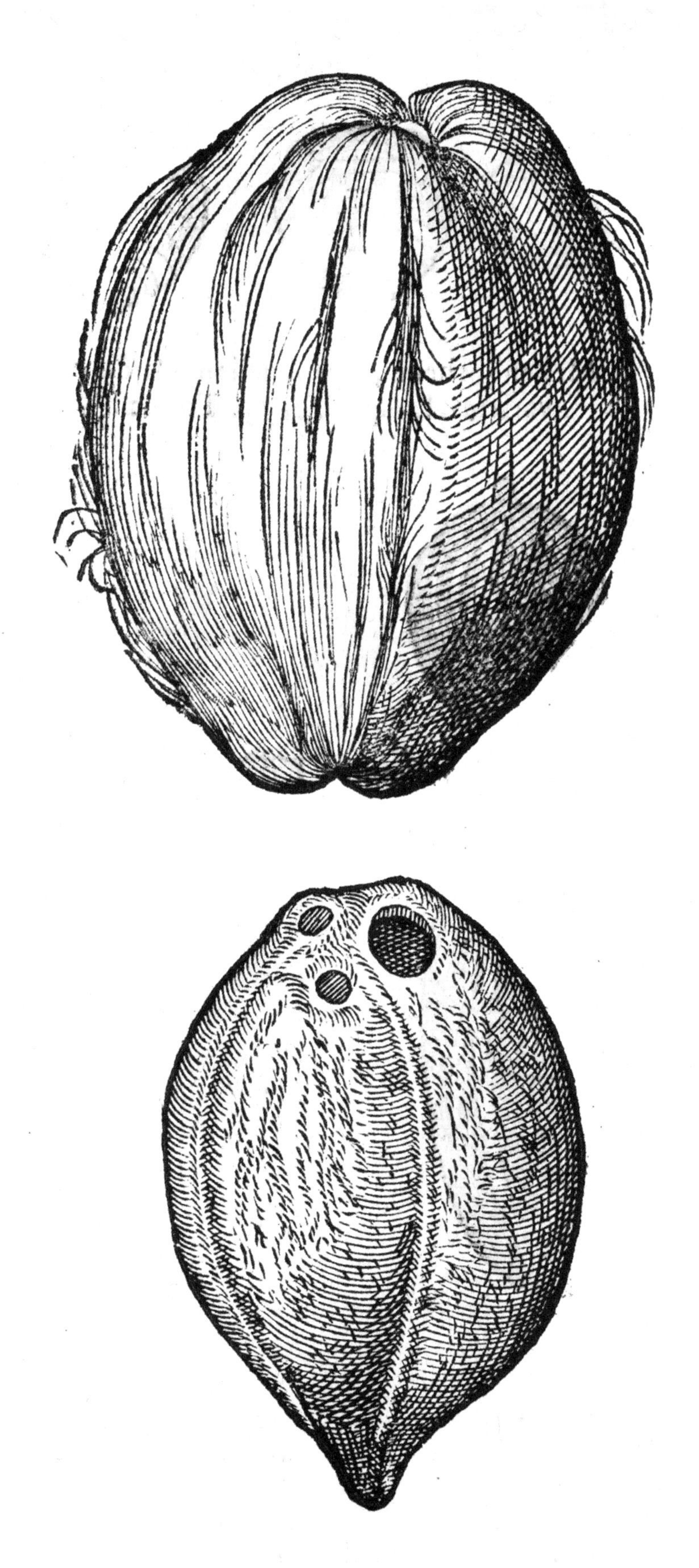

In the mid-nineteenth century Georg Hartwig summarized coconut distribution thus:

> essentially littoral, this noble palm requires an atmosphere damp with the spray and moisture of the sea to acquire its full stateliness of growth ... loves to bend over the rolling surf, and to drop its fruits into the tidal wave. Wafted by the winds and currents over the sea, the nuts float along without losing their germinating power ... the cocoa-palm has spread its wide domain ... throughout the whole extent of the tropical zone. It waves its graceful fronds over the emerald islands of the Pacific, fringes the West Indian shores, and from the Philippines to Madagascar crowns the atolls, or girds the sea-borde of the Indian Ocean.[395]

We are fascinated by patterns we observe in the natural world, and seek explanations for them. Hartwig's florid description masks centuries of rich, sometimes vituperative, academic debate over where coconut originated and how and when the ends of the coconut girdle around the Earth were joined.

There were no coconut palms in West Africa, the Caribbean or the east coast of the Americas before the voyages of Vasco da Gama and Columbus. Trade patterns and human migration reveal that Arab traders and sixteenth-century Iberian sailors are likely to have moved coconuts from South and South East Asia to Africa and then across the Atlantic to the Americas.[396] The Treaty of Tordesillas (1494), which divided the non-European world between Lisbon and Madrid, prevented Spaniards from travelling to Asia (see NUTMEG, page 137), so it is unlikely that any saw living coconuts in the early sixteenth century. The discovery of coconuts along the west coast of Panama by sixteenth-century sailors has therefore provoked centuries of discussion as to their origins. The favoured view today is that this final link in the coconut girdle was created by Iberian sailors moving coconuts from the Philippines to the Americas.[397] But where did the girdle start? Two diametrically opposed origins have been proposed: America and

Asia. Both suggestions have problems. In Asia there is a large degree of coconut diversity and evidence of millennia of human use – but no wild relatives.[398] In America there are close coconut relatives but no evidence that coconuts are indigenous.[399] These problems have led to the intriguing suggestion that coconuts originated on and were dispersed by populating dynamic coral atolls in the Pacific.[400]

During the nineteenth century coconut oil and coir became items of international commerce. At least one author attributes the change in status to two serendipitous Victorian social events: Queen Victoria's wedding in 1840, and the christening two years later of her eldest son, Prince Albert Edward, eventually to be crowned King Edward VII.[401] For the wedding an enterprising candle-maker made high-quality, cheap candles for the occasion from coconut oil. For the latter event, coir matting covered the floor of St George's Chapel in Windsor Castle. By the time of the Great Exhibition in 1851 coir mats and ropes, and coconut candles, were all the rage and making Victorian businessmen rich; most of the raw materials conveniently grew in profusion in imperial possessions, particularly Sri Lanka and India.[402] At much the same time, a derivative of coconut fat, glycerine, acquired strategic importance in a quite difference sphere, as Alfred Nobel introduced the world to his nitroglycerine-based invention: dynamite.

It is only in the past few decades, as other high explosives became available and health concerns over saturated coconut fat developed, that the market in the West for coconut fat has declined.[403] Despite the ubiquity of coconuts and their importance to human cultures across the globe, we still have much to learn about one of our most useful plants.

The Cock

Rice

Oryza sativa L.; Poaceae

FEEDING BILLIONS

Rice feeds more people than any other plant on the planet. For at least 9,000 years Asians have depended on rice, have modified rice to their lives and in turn been modified by the demands of rice. Different types of rice are adapted to the multiplicity of Asian environments ranging from the tropics to the temperate latitudes of northern Japan, from the Cambodian lowlands to the Himalayas, and from the deep waters of Bangladesh to the dry uplands of Nepal. Consequently, there is tremendous variation in rice characteristics, and in the everyday uses and significance of rice in Asian societies. However, it is only in the last 250 years that rice has become a familiar part of Western diets.

The history of rice is the history of contacts between Eastern and Western civilization. Alexander the Great's foray into India occasioned the Greek Theophrastus to describe rice for his audience as early as the fourth century BCE.[404] With the rise of the Islamic world, rice culture extended to the southern shores of the Caspian Sea, and by the tenth century the Moors had taken rice to the Iberian Peninsula. The fifteenth century saw rice spread through Italy and France, and to the Americas. Rice is not native to the Americas but it became a staple; today, it is hardly imaginable to have a meal in Latin America without some combination of Old World rice and New World beans. As the colonial south of the United States started to benefit from rice cultivation and the slave trade, the high prices paid for slaves familiar with West African rice cultivation has led some academics to place black slaves at the centre of rice culture in colonial and post-colonial America.[405] By the end of the nineteenth century, despite its eastern origin, rice was part of the daily lives of people in the West.

Rice is an annual grass of two main types. The *indica* type has non-sticky, long grains and is adapted to warm, temperate climates. The other, the *japonica* type, has sticky, short grains and is adapted to damp tropical climates. Over thousands of years people have participated in vast communal engineering feats to terrace steep hillsides, dam fields and channel water for paddy fields. Although not essential for rice cultivation, flooding reduces weeding and encourages the fertilization of fields by the nitrogen-fixing aquatic fern *Azolla*. The number and area of paddy fields has increased with rice production to such an extent that today they are responsible for approximately 48 per cent of the annual emissions of the greenhouse gases from agricultural land.[406]

Trying to understand the origins of domesticated rice is a challenge, because of the vast areas of potential rice cultivation, the few sites where archaeological remains might be preserved, and the difficulties of distinguishing wild and cultivated rices. Available evidence suggests the *indica* and *japonica* types diverged well before rice was domesticated; today they are so different they cannot interbreed. However, both types do interbreed with rice's wild ancestor *Oryza rufipogon*, a perennial species distributed across Asia. The scientific debate continues about whether domestication happened once or on multiple occasions.[407]

Rice in Africa has had a different history from that in Asia. African rice, *Oryza glaberrima*, was domesticated from *Oryza barthii* about 3,500 years ago in the Upper Niger Delta. As with Asian rice, there is an unresolved debate over the single or multiple origins of African rice.[408] African rice has a hard grain and is resistant to diseases and pests, but it has low yields and is difficult to harvest. Following the introduction of Asian rice into the range of African rice, the two coexisted side by side until consumers' preferences meant it was more economic to grow Asian than African rice.[409]

Rice grains are natural packets of food reserves, and for thousands of years they have ensured crops are produced annually; they have passed their genes from generation to generation and migrated with humans, adapting as they went. Scientists and plant breeders have exploited these characteristics to store genetic potential. It is believed that most natural genetic variation in Asian rice is captured in seed banks. We rely on this inherited genetic complexity for breeding new rice cultivars today, and will do so into the foreseeable future.

Traditional approaches to plant breeding involve manipulation of the frequencies of genes. This strategy has enabled humans to feed themselves, so far. However, the challenges posed by climate change, increasing human population and habitat loss have led breeders to consider approaches that involve moving specific genes into crops, the re-creation of crop plants from their component species or manipulation of specific genes in existing crops. To make the necessary step changes in rice productivity it has been suggested that rice photosynthesis must be made more efficient, that the entire biochemistry of photosynthesis should be re-engineered.[410]

Rice is nutritionally incomplete. Cleaned rice grains, for example, are deficient in vitamins such as A and B1, and minerals such as iron. Consequently, vitamin-deficiency diseases are frequent in people with rice-dominated diets. Vitamin A deficiency has enormous public health consequences; between 250,000 and 500,000 vitamin A-deficient children go blind each year, and half of them die within twelve months. Strategies for alleviating vitamin A deficiency have met with varying degrees of success, but one proposal was the development of Golden Rice. The first variety of Golden Rice had high levels of a vitamin A precursor, which was achieved by introducing three new genes, two from daffodil and one from a bacterium. A second Golden Rice variety used a maize gene instead of the daffodil genes.[411] Golden Rice is not the complete solution to vitamin A deficiency worldwide, but both Golden Rice varieties are now being used in Asian rice breeding programmes. Whether scientific agriculture can ensure rice production keeps pace with human population growth remains to be seen.

Critics of modern crop breeding have argued that people's traditional knowledge has been exploited and their biological heritage stolen in what has been termed biopiracy. However, feeding the growing human population, which has changing expectations, is the major challenge for all societies. Rice is at the sharp end of the biopiracy debate since it is consumed by so many poor people. Scientific developments in rice breeding, and other crops, must be accompanied by acceptable social changes. To be most effective, increasing food production must be associated with reductions in both population growth and overall consumption; there must be fewer, slimmer people on the planet.

Tea

Camellia sinensis (L.) Kuntze; Theaceae

A BRITISH STEREOTYPE

Few plants bear the mark of imperialism as strongly as the Asian evergreen shrub, tea. Indeed, tea is virtually inextricably linked to the histories of the Chinese and British empires, particular during the nineteenth century. Western cultures began to experiment with this new drink in the sixteenth century, but in the East people had used it for centuries, albeit medicinally. Eastern origins of the adoption of tea are lost in the fog of legend, although we know that the Chinese were drinking tea by 350 BCE. Furthermore, the Chinese had created well-guarded, sophisticated methods of processing it – methods that remained unknown to 'foreign devils' until the mid-nineteenth century. Ignorance of how tea was processed meant arguments over the identity of the species producing black and green tea filled the botanical literature; the eighteenth-century botanical celebrity Carolus Linnaeus was one of many convinced that two different species were involved.

By the late seventeenth century tea was an expensive, often adulterated, luxury transported to Europe overland or by sea from China. During the following century tea began to transform international commerce as the East India companies formed by European countries entered into complex trade relationships with China. China wanted nothing Europeans could offer for their tea, except precious metals. The annual silver tribute from Europe to China was enormous. In Britain tea became an eighteenth-century social fad among the well-to-do, where a refined, often pretentious and proscriptive, culture developed associated with its consumption. Naturally, governments took advantage of the tea trade, taxing it heavily, which encouraged smuggling. The so-called Boston Tea Party in 1773 was not all about tea, but a manifestation of the disquiet with taxes levied by London in the American colonies. 'No taxation without representation' cried the revolutionaries, and the tax on tea and other imports was a contributing factor to the American War of Independence (1775–82).

The pepper trade had diminished the exchequer of Ancient Rome (see PEPPER, page 115); European governments were perhaps fearful of a repeat with the tea trade. Consequently, they struggled to find ways of subverting Chinese control of tea production. China, a vast country about which Europeans knew very little, placed considerable restrictions on the movements of foreigners within her borders. From the 1740s until the end of his life, Linnaeus tried to import tea plants into Sweden.[412] He was convinced Europe could become self-sufficient in tea, and that Sweden could lead the way. Five of his students, who set out on plant-collecting expeditions to China, had orders to collect tea plants and return them to Sweden; they all failed. Linnaeus eventually got living tea plants in 1763; unsurprisingly, these failed to grow in the Swedish climate. In the early nineteenth century the British made limited efforts to introduce tea to their Indian colonies and secure a tea supply for a British population that was increasingly seeing tea as essential to life. Tea was taken up by nonconformists and the temperance movement as a refreshing, acceptable alternative to 'the demon drink', and during the British Industrial Revolution the urban poor appeared to survive on little more than bread, sugar and tea, leading to criticisms that they were becoming dependent on tea.[413]

In the mid-nineteenth century the American lawyer Junius Smith, best known for championing transatlantic steamships, exhorted American farmers to free tea from the Chinese 'bondage of intellectual servitude'.[414] Like Linnaeus a century before, he was convinced tea would flourish in his homeland. He arrogantly dismissed Chinese understanding of tea processing, assured that entrepreneurs in the

Thea frutex Bontii. Chaa.

1. Blüthe
2. 3. Frucht
4. offene Frücht
5. Saame

Thee peco Thee bohe. Rother Thee.

United States, 'Middle Kingdom of the World', would do better. Like Linnaeus, Smith was wrong.

Tea has never become a major crop of the Americas, despite various dalliances with its cultivation in the nineteenth century, most notably in Brazil. Tea from plantations in the Rio de Janeiro Botanic Garden in the 1840s was damned with faint praise by the Scottish botanist George Gardner: 'in appearance [it] is scarcely to be distinguished from that of Chinese manufacture, but the flavour is inferior, having more of an herby taste.'[415] Even had the quality been good, it could never have competed with the volume that was produced by the next phase in the global ascendancy of tea.

At about the same time as Smith was pontificating, the British East India Company commissioned another Scottish botanist, Robert Fortune, to obtain living tea plants.[416] Fortune was an expert on Chinese plants, and familiar with Chinese customs. He often travelled covertly, adopting Chinese dress, and moved into parts of China unknown to westerners. It was Fortune who ascertained that green and black teas were produced by the same plant, and that the differences were in the processing.[417] When he succeeded in shipping 2,000 tea plants and 17,000 tea seeds to India from Hong Kong – a new colony added to the British Empire after the First Opium War (see OPIUM POPPY, page 33) – he also had the wit to ship Chinese workers who had proven knowledge of tea cultivation and processing.

In addition to caffeine, tea plants contain complex mixtures of chemicals called polyphenols. The flavour of tea depends on where tea plants grow, the quality of harvested leaves and how the polyphenols are modified during processing,[418] so effective processing of the fresh leaves is important. Wilting, bruising, oxidizing, drying and curing all change tea-leaf chemistry, and contribute to the familiar classes of tea such as green, oolong and black tea.

Fortune's Chinese tea plants were established in northern India, together with the native Indian tea plants discovered in Assam.[419] From there tea plantations spread across the British Empire, guaranteeing a cheap supply of the beverage that has fuelled the Commonwealth ever since. The perceived need for speed in the race to return tea from the colonies to London produced the sleek lines of the tea clippers, constructed from timbers such as elm and teak.

Fortune was successful in his bid to acquire tea plants because he was an excellent botanist, had intimate practical knowledge of Chinese culture and was aware that the technology of tea processing was as important as the plants themselves. He also made use of novel methods to transport his precious plants. For centuries, transporting living plants was not for the faint-hearted, since potted plants on board ship require constant attention, a supply of fresh water and protection from the elements. Then in 1823 Nathaniel Ward, a London doctor, made a transportation breakthrough: the Wardian Case. The closed glazed box he invented, rather similar to a portable greenhouse, protected growing plants from unfavourable conditions and opened up myriad new possibilities for plant hunters and horticulturalists.[420]

Despite the global flood of coffee shops, tea remains the world's most important caffeine-containing beverage. Tea and coffee cultures split people and nations. Tea drinking is often seen as rather conventional, associated with moderation and temperance, refined afternoons and the English stereotype. In contrast, coffee drinking is perceived as bohemian, associated with excess and indulgence, early mornings and late nights and the American stereotype. Over the past decade, health benefits have also been attributed to tea, bringing tea consumption back to the Chinese practice of more than 1,500 years ago.

a. *Iacobæa vulgaris major, Iacobæe.*
b. *Iacobæa vulgaris laciniata floribus albicantibus.*

Ragwort

Jacobaea vulgaris Gaertn.f.; Asteraceae

UNINTENDED CONSEQUENCES

Crops and the plants that associate with them are the footprints of human colonization. Major food plants have moved, under the pampering hand of humans, out of their cradles of origin to colonize vast continental areas. Along with these migrations, other plants have been moved accidentally, taking full advantage of opportunities offered by new man-made environments and humans' abilities to cross biologically insurmountable barriers. Some of these plants have earned the appellation 'weed'. Weeds are botanical camp followers, opportunists capable of colonizing the habitats created by humans and hitching lifts as humans or their animals move. Definitions of weeds range from those plants 'whose virtues have not yet been discovered', by the weed enthusiast,[421] to 'a plant in the wrong place' by the more practically minded. Weeds have short life cycles, are good at getting around and exploiting disturbed areas, and typically produce large numbers of seeds or fragment their own bodies. Ragwort has all these characteristics.

Ragwort, with its flat heads of yellow daisy-like collections of flowers, is native to Eurasia and for decades was treated as a member of the enormous genus *Senecio*. However, DNA analyses have led to it, along with its close relatives, being transferred to the genus *Jacobaea*,[422] a name which reflects the former belief that ragwort starts to flower around St James's day (25 July). Over the last 150 years, as European empires moved livestock across temperate North America, Australia and New Zealand, it has become a familiar, and serious, pastureland weed. In Britain, ragwort regularly attracts emotive headlines as a horse destroyer. Less parochially, but more importantly, it is responsible for the annual loss of hundreds of millions of pounds of pasture productivity globally.

The current relationship between humans and ragwort is but the latest stage in the story of how ragwort has taken advantage of man since the Neolithic Age. In Europe ragwort is a native of highly disturbed, well-drained soils, such as sand dunes.[423] As humans changed the landscape following the end of the last Ice Age, they cleared forests to create agricultural land, exposing the soil to regular disturbance by wind, water and plough, so inadvertently enticing ragwort to expand out of its marginal native habitats. As livestock densities increased, their pastureland was likely to be regularly overgrazed, providing opportunities for ragwort seeds to establish themselves free from competition from other species, a pattern repeated by many familiar agricultural weeds.

Individual ragwort plants produce thousands of wind-dispersed seeds (technically single-seeded fruits), and they also readily regenerate from stem or root fragments,[424] so it is easy to see how suitable habitats may rapidly become overrun with ragwort, and effective control may become impossible. Heavy ragwort infestations reduce pasture productivity through competition with grasses and other herbs. Furthermore, ragwort is one of a relatively small number of plants containing a class of compounds called pyrrolizidine alkaloids, some of which are highly toxic to grazing animals[425] – hence the dramatic, frequently mythologized, reputation of ragwort as a livestock killer. Livestock generally avoid ragwort because of the toxins, although they will eat it if it has been dried along with hay,[426] and livestock poisoning can have an economic impact[427] as ragwort densities increase. However, given the economic and emotional reactions, it is unsurprising that obtaining objective evidence of the numbers of animals specifically killed by ragwort is difficult.

Legislation to control the spread of ragwort, such as the UK's Weeds Act (1959) and the later Ragwort Control Act (2003) in England, may seem a vain response to a plant so well adapted to humans' lifestyles.

Yet we may take advantage of those insects that are adapted to using ragwort in order to control it. Some of these insects – for example, the caterpillar of the cinnabar moth, and larvae of the ragwort flea beetle and the ragwort seed fly – are effective biological control agents in parts of ragwort's introduced range.[428] But effective biological control will only reduce ragwort populations to levels where there are negligible economic effects. Additional controls, such as pasture improvement, reduce opportunities for ragwort seedlings to establish themselves, while regular cropping destroys adult plants and interferes with seed dispersal.

As one might expect for such a reviled plant, ragwort is known by hundreds of different local names. One name, staggerwort, appears to describe the symptoms of livestock suffering ragwort poisoning, although at least one source recommended ragwort as a *remedy* for staggers in horses![429] Rather than seeing it as a ravager of landscapes and destroyer of livestock, some have a more benign view of ragwort. Under its Manx name, *cushag*, ragwort is the national flower of the Isle of Man, while the poet John Clare wrote 'The Ragwort' about the plant's beauty; it also attracted the attention of flower fairy author Cicely Mary Barker in 1925.

As a weed, ragwort represents those plants that have exploited our foibles. It has taken the opportunities to spread out of its limited native habitats as Western civilization expanded and manipulated its surroundings. As an unforeseen consequence of civilization, it has crossed continents in the feed and bowels of livestock and the slipstreams of vehicles, while poor land management has provided ideal conditions for its establishment. Culturally, it has attracted both praise and opprobrium, although not in equal measure. As horses lost their role as primary means of transport and rural power, the significance of ragwort has shifted from an economic to an animal welfare problem.

Bañana. 2.
Mauze.
Paradiß-Feigen.

Banana

Musa × paradisiaca L.; Musaceae

FRUIT ON A KNIFE-EDGE

In 2019 the European media became excited by one of their periodic food scares. This scare was about neither contamination nor carcinogens but about the discovery of a fungal pathogen of banana in Colombia. Bananas are the fourth most important food staple in the world: about 135 million tonnes are consumed annually, and they are one of the top ten most-traded agricultural commodities. For North American and European consumers the demise of the banana would be the loss of their favourite fruit, but the real threat was to the countries that rely on the export of bananas and to the lives of more than half a billion people whose staple starch source is plantains.

Although people may refer casually to 'banana trees', bananas and plantains are in fact very large herbs. They are native to the Malay–Indonesian region, but are now planted throughout the tropics and subtropics. When Linnaeus named bananas *Musa sapientum* and plantains *Musa paradisiaca* he was reflecting a widely accepted belief that the banana was the Tree of Knowledge in the Garden of Eden. The English herbalist John Gerard called them 'Adam's Apples' and reported the belief that cutting a banana revealed the image of Christ on the cross. However, once he had investigated pickled specimens for himself, he wrote: 'the crosse I might perceive, as the forme of a Spread Egle in the roote of Ferne, but the man I leave to be sought for by those that have better eies and judgement then my selfe.'[430] Generally, bananas are sweet and eaten raw, while plantains are starchy and eaten cooked, but botanically the distinction between the two is entirely

Banana. 1.

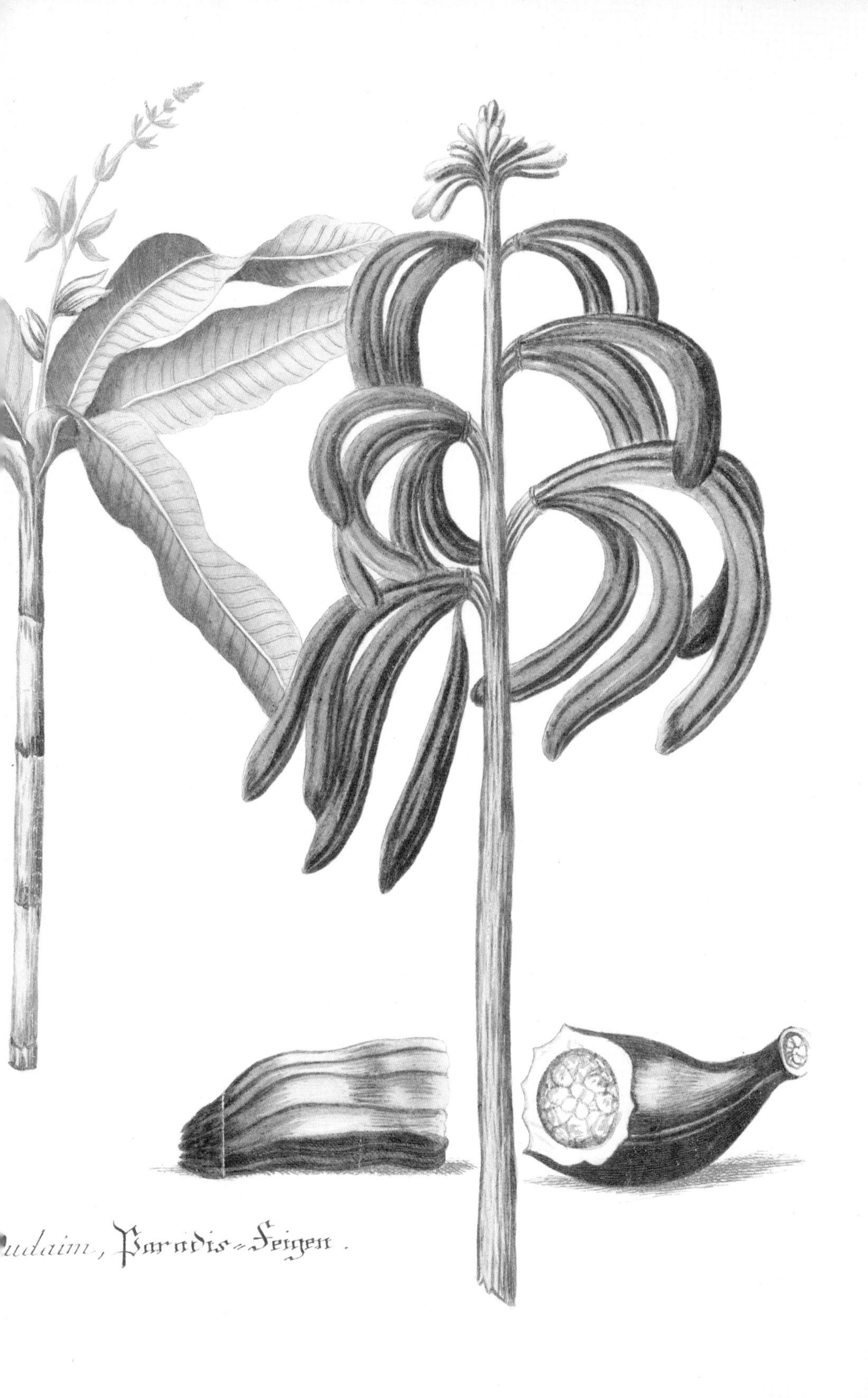
udaim, Paradis-Feigen.

artificial and now that we understand more about banana evolution we treat them as one.

Bananas are latecomers in the history of Western civilization's relationship with plants, but evidence from Papua New Guinea indicates banana cultivation may be as much as 10,000 years old, with its domestication happening independently across South East Asia and parts of Africa.[431] Most cultivated bananas and plantains are derived from complex ancient crosses between two wild species, *Musa acuminata* and *Musa balbisiana*.[432] Fruits of the wild species have seeds but those of cultivated bananas and plantains are seedless.

Bananas steadily made their way out of South East Asia and, in the wake of Islam, spread through the Near East and North Africa into the Iberian Peninsula; by the ninth century bananas were familiar motifs in Islamic art and literature.[433] During the sixteenth century the Portuguese introduced bananas to the Americas, via West Africa and the Canary Islands, but they remained novelties in the capitals of Western Europe. The artist who prepared the frontispiece of John Parkinson's *Theatrum Botanicum* (1640) included a banana plant but had evidently never seen one. Rare, but not altogether unknown: as early as 1633 the apothecary Thomas Johnson waxed lyrical over the novelty of being given a bunch of green bananas, which he displayed in his London shop and observed ripening over about six weeks.[434] By 1736 Linnaeus had coaxed a banana plant, introduced to the Netherlands from Suriname, to flower in the Dutch financier George Clifford's de Hartecamp garden; the first banana to flower in Europe.[435] For a long time, however, bananas remained rare exotics – before they could become commonplace in Western markets it was necessary to have rapid transport systems and mechanisms, such as refrigeration, to slow ripening. Only in the last eight decades have these conditions been fulfilled, and bananas become the tropical cliché, slapstick prop and supermarket mainstay with which we are familiar.

There are thousands of banana cultivars, naturally occurring mutants selected for vigour, yield, hardiness, fruit quality and, importantly, seedlessness, divided into several distinct groups. Despite all this apparent diversity, only two types have found favour in international trade, 'Gros Michel' and 'Cavendish', both discovered in the early nineteenth century. 'Cavendish' was named in honour of William Cavendish, the sixth Duke of Devonshire, in whose glasshouses at Chatsworth House the commercial potential of the cultivar was developed.[436] The 'Cavendish' may be popular, but it is not the most flavoursome banana when compared to local cultivars; in Brazil it is disparagingly called *banana d'água* (water banana).

Since banana cultivars are seedless they cannot reproduce without human help, so once in cultivation they must be multiplied by vegetative propagation. In this lies their potential downfall. The paucity of cultivars in commercial cultivation, combined with vegetative propagation, which does not allow for the natural variation that seed propagation encourages, means that banana production is particularly vulnerable to disease outbreaks. Commercial production of 'Gros Michel' has been uneconomic since the 1950s because of Panama disease, one of the most destructive plant diseases known,[437] and one that cannot be controlled using fungicides or fumigants.

Now the 'Cavendish' banana is threatened by a new strain of Panama disease and also by another fungal pathogen that causes black leaf streak disease, and loss of up to half the crop.[438] Other than prevention, the only viable option to maintain productivity is cultivar replacement.[439] Within bananas as a whole there is plenty of fungal resistance; the solution will lie in getting this resistance into commercial bananas. There is hope that scientists will find the key through further studies of banana's DNA, which was completely sequenced in 2012.[440] The cataclysm predicted in 2003 is unlikely, but such reports highlight the vulnerability of crops to acts of gods and the need for vigilance.

During the 1970s and 1980s bananas became symbolic political weapons as the United States and Soviet Union fought the cold war in banana-based economies, the so-called Banana Republics. Now, bananas are symbolic in a different struggle: the vast profits made by Western food retailers through the international banana trade have become a focus for campaigners concerned with the ethics and environmental costs of food production.

Rubber

Hevea brasiliensis Müll.Arg.; Euphorbiaceae

BALL GAMES TO CONDOMS

It was an American ball game that gave Western civilization a vital component of the industrial age. When sixteenth-century Europeans began exploring the continent they had recently rediscovered, they encountered much completely outside their experience, including an elastic waterproof material that bounced. Although a novelty to Europeans, rubber had been used for at least 2,500 years to make the balls for the Mesoamerican ritual and sport *ulama*.[441]

Discovery of two chemical treatments transformed rubber from a curiosity with few practical uses to an adaptable and soon indispensable raw material. First, it was found in the eighteenth century that rubber was soluble in organic solvents manufactured from coal. This meant rubber could be moulded into different shapes or used to impregnate fabrics for waterproof clothing. Second, nineteenth-century scientists discovered the elastic properties of rubber could be altered by treating it with sulphur. Vulcanization, as it was dubbed, meant rubber could be used to make machinery washers and belts and, importantly, pneumatic tyres for bicycles and motor cars. The rubber age had started, but, in the killing fields of commerce in Amazonia and the Belgian Congo, rubber extraction would cost millions of lives.

Referring to rubber in the singular is not quite correct, as many different, unrelated plants produce natural rubbers, which are polymers of colourless, volatile hydrocarbons called isoprenes. What rubber-producing plants have in common is a white or coloured latex found in special latex vessels, which are part of the plant's defence system against herbivores or disease attack. Many gardeners will be familiar with the milky, sticky sap exuded by, for example, euphorbias and figs – the Mesoamericans would use latex from a type of fig when making their *ulama* balls.[442] By carefully cutting a plant's bark, latex vessels can be broken and the latex repeatedly tapped without killing the plant. Traditionally, chewing gum was made from rubber extracted from Mesoamerican manilkara trees. Gutta-percha, the rubber that heralded the age of electrical insulation, dental fillings and revolutionary golf balls, was extracted from a South East Asian tree. Although many natural rubbers have been replaced by synthetic alternatives, natural rubbers remain important, especially Pará rubber.

In 1869 James Collins identified Pará rubber, extracted from a tree native to the Brazilian Amazon, as an ideal rubber source; a position it has maintained to the present day. Pará rubber was so well suited because of its properties but also because of the prospect that British imperial control over supply could be achieved by growing it in South East Asia. The senior civil servant Clement Markham, who masterminded the introduction of cinchona to India from South America (see QUININE, page 163), took up the cause. However, in the corridors of Whitehall nobody understood Pará rubber's biology, had any seed or knew how to extract rubber, even if the tree could be grown. The British adventurer and aspiring plantation owner Henry Wickham filled this gap.

Wickham knew the rubber tree in Amazonia; he had even seen rubber being harvested. Joseph Dalton Hooker, the director of the Royal Botanic Gardens at Kew, decided to commission Wickham to collect seeds, germinate them at Kew and then ship the seedlings to the South East Asian colonies. Wickham arrived, unannounced, at Hooker's house at Kew in the early hours of 14 June 1876 with 70,000 Pará rubber seeds that he had brought back from the River Tapajos, Brazil. After this, events moved fast. The next day, all 70,000 seeds were planted at Kew; three weeks later, 2,700 seeds had germinated. In August, 1,919 seedlings (records are very precise) were sent to Sri

Lanka, and the following month 1,700 seedlings arrived safely at the Henarathgoda Gardens, north-east of Columbo. Henarathgoda had been established to nurture the prized saplings; there are even claims that the stump of one of the original trees survives to this day in the Gardens.

Within weeks of Wickham's return, another Kew emissary, the gardener Robert Cross, was making his way to Brazil to collect rubber tree seed. Cross returned to Kew in November 1876, and 100 Pará rubber plants were subsequently propagated from his seed and sent to Sri Lanka. The following June, twenty-two plants from Cross's collection were sent on to Singapore. Henry Ridley, father of the South East Asian rubber industry, always claimed that most Pará rubber trees grown in South East Asia were propagated from Cross's seedlings, but the controversy over whether the plants were propagated from seedlings supplied by Wickham or Cross remains.

As the British were establishing Pará rubber plantations in their South East Asian territories, the Amazonian rubber boom was gathering pace. Brazil, and states in north-western South America, had a virtual monopoly on rubber supply and, as workers were being exploited, the so-called rubber barons became extraordinarily wealthy. So confident were people that rubber money would provide the base for a rich and cultured population that the construction of an anachronistic Renaissance-style opera house in Manaus, a city halfway up the River Amazon, was funded. By 1912, just fifteen years after the first performance in the opera house, the Amazonian rubber balloon had been punctured. *In The Lost World*, published in the same year, Arthur Conan Doyle included the breaking of the rubber barons as part of Lord John Roxton's backstory. One cause of the collapse was that rubber could be extracted more cheaply and efficiently from the British-controlled plantations in South East Asia, and one of the reasons for the success of the Asian plantations was that trees did not suffer from the pests and diseases there. Today, tens of millions of tonnes of rubber are produced annually, of which about 40 per cent is from natural sources. Of these natural sources, the vast majority is produced by Pará rubber plantations in former British colonies in South East Asia.

Understanding the structure of natural rubbers means we can manipulate their physical properties and synthesize them to our own requirements. The single most important use of natural rubber is in the manufacture of tyres and tubes, particularly for aircraft. On 28 January 1986, failure of a rubber O-ring led to an explosion minutes into Space Shuttle *Challenger*'s flight. This event, witnessed on television by millions, showed in dramatic fashion the potential fallibility of civilization's natural products and the dangers of exploration. Yet, through its use as the raw material for condoms, Pará rubber has positive social effects on family planning and the control of sexually transmitted diseases.

The story of rubber, including Kew's role in introducing it to Asia and the dubious legality of Wickham's involvement, is often considered a tangible example of where biopiracy can lead, and the questions their actions raise continue to be relevant.[443] Almost exactly a century after Wickham's original exploits, the rights and responsibilities attached to Pará rubber made another brief appearance on the world stage. In 1988 Chico Mendes, an Amazonian rubber tapper closely associated with movements to conserve the Amazonian rainforest, was shot.[444] His murder threw a spotlight on conflicts over land rights and the place of people in exploiting natural resources.

Chrysanthemum Indicum flore et semine maximum.
Sonnen-blum.

Sunflower

Helianthus annuus L.; Asteraceae

CURIOSITY TO COMMODITY

Our interactions with plants change over time, and diverse cultures view them in different ways. In its native habitat, Native Americans saw the sunflower as a major food and medicine plant, one that also usefully supplied fibre and dye and was even a source of musical instruments and bird snares.[445] Yet, in Europe, following their introduction from the Americas in the sixteenth century, sunflowers were little more than garden novelties. Their towering size was surely one of their attractions, although in 1597 Gerard appeared disappointed the sunflowers he grew in his Holborn garden were *only* 4.3 metres tall; those of his European competitors reached 7.3 metres.[446]

Sunflowers proved popular, though, and by the start of the seventeenth century the apothecary John Parkinson could claim 'this goodly and stately plant, wherewith every one is now adayes familiar'. Parkinson went on to assert 'there is no use … in Physicke with us, but that sometimes the heads of the Sunne Flower are dressed, and eaten as Hartichokes are, and are accounted of some to be good meate, but they are too strong for my taste.'[447] Confusingly, while the globe artichoke to which Parkinson was comparing the sunflower head is from a different genus altogether, the Jerusalem artichoke *is* a sunflower (*Helianthus tuberosus*), introduced from North America in the early seventeenth century. The name comes not from any link with the Holy Land, but is a mangling of the Italian *girasole* (turning with the sun) and a perceived similarity in flavour between its tubers and globe artichoke flesh.[448] John Goodyear, writing on this new addition to the culinary canon, passed his opinion quite forcibly on its shortcomings:

> these roots are dressed divers waies; ... but in my judgement, which way soever they be drest and eaten they stirre and cause a filthie loathsome stinking winde within the bodie, thereby causing the belly to bee pained and tormented, and are a meate more fit for swine, than men.[449]

The 'stinking winde' is produced by gut bacteria as they break down the sunflower storage polysaccharide, inulin, which, unlike starch, we cannot digest directly.

The massive flower heads of sunflowers, balanced on a single stem, demand attention. In fact, a sunflower head is not a single flower but comprises hundreds of tiny flowers. Those around the outside of the head (ray florets) have large, bilaterally symmetrical groups of petals and are sterile. Those on the inside of the head (tube florets) have symmetrical corollas and are fertile. Each tube floret produces one dry fruit, containing an oil-rich seed, which the English herbalist John Gerard described as 'set as though a cunning workeman had of purpose placed them in very good order'.[450]

Gerard's observation was more than a casual description. Careful examination of a sunflower head reveals that the individual seeds within their encasing papery 'fruits' form two series of spirals; one set running clockwise and the other anticlockwise. The artists contributing to Gerard's *Herball* (1597), Besler's *Hortus Eystettensis* (1613) or Parkinson's *Paradisi in sole* (1629) baulked at showing such intricate patterns, but we can all see them once we know to look.

The spirals are laid out with mathematical precision, and are a natural example of the Fibonacci series. Mathematics is a civilizing discipline, and the mathematics of the sunflower head has attracted attention since at least the mid-eighteenth century. The French mathematician Edouard Lucas named the Fibonacci series in honour of the thirteenth-century Italian mathematician Leonardo Bonacci of Pisa, known as Fibonacci. In his *Liber abaci* (1202), Fibonacci identified a number sequence where each successive number is the sum of the previous two (1, 1, 2, 3, 5, 8, etc.).[451] This series has come to be harnessed in the fields of computing, finance, gambling and many others for predictive research, and the pleasing rhythms of the sequence recur in music, art and architecture.

This display of the Fibonacci series is a consequence of how a plant grows, and is frequently found in the way structures are arranged to fill limited space. Packing in the sunflower head is maximized by having adjacent fruits arranged at precisely 137.5 degrees to each other.[452] This economy of growth pattern, and the resulting mathematical spirals, can also be seen in other plants, including pine cones and pineapples.

Current evidence shows that the sunflower was domesticated once, about five millennia ago; it is one of the few major crops to have been domesticated first in North America. However, active discussion surrounds the interpretation of archaeological, linguistic and literary evidence over the extent to which sunflower was part of daily and religious life in pre-Columbian Mexico.[453]

Interest in sunflower seeds as a source of oil was generally slow across Europe, except in Russia. Perhaps because of edicts by the Russian Orthodox Church that sunflowers were not prohibited during Lent, the Russians became major developers of industrial sunflower oil production. By 1830 there were hundreds of thousands of hectares of sunflowers in commercial production,[454] and Russian plant breeders were selecting sunflowers with seeds most suitable for oil extraction and direct consumption. By the late nineteenth century sunflower seeds were being returned to North America, probably in the baggage of Russian immigrants, not as low-yielding botanical novelties but as highly productive cultivars, such as 'Mammoth Russian'. During the early twentieth century the Soviet plant breeder Vasilii Stephanovich Pustovoit raised the levels of oil in sunflower seeds dramatically. Sunflowers became part of cold war industrial espionage and biopiracy as Soviet seed became part of today's global sunflower oil industry.[455] In 2022 more than 54 million tonnes of sunflower seeds were produced, of which over 50 per cent were grown in Ukraine and the Russian Federation.

Seed oil yield was identified as an early goal of sunflower breeders, but other characteristics are also important. Careful breeding has allowed different requirements to be met, taking advantage of the sunflower's natural variability. Competitive horticulturalists, for example, may regard large, brightly coloured heads on tall plants as the premium feature, whereas for commercial farmers tall plants with large heads are likely to fall over and reduce yield, and their preferred characteristic would be nodding heads to help protect the fruits from birds.

Sunflower oil, which accounts for most of the crop's commercial worth, is highly adaptable. It is valuable because it is edible, has little taste, a light colour, high levels of unsaturated fatty acids and is perceived as a healthy alternative to saturated animal fats. Sunflower oil is particularly popular in Europe and the Russian Federation, where it is a cheap substitute for olive oil, and is used for cooking, margarine production and even biodiesel manufacture. Residues that remain following seed oil extraction provide a high-protein animal feed, often being used as substitutes for soya. The Russian invasion of Ukraine in February 2022 and the subsequent war in the country continue to affect global prices of sunflower oil and the myriad products containing it.

For much of their history in Europe sunflowers were little more than eye-catching flowers that lived up to their name. Their size and brilliance have always captured attention. In 1987 the most expensive painting in the world (briefly, for such figures are quickly superseded in the art world) was one of Vincent Van Gogh's sunflower paintings, which sold for nearly $40 million. The true value of sunflowers, however, lies in their worth as an international commodity and the multiple, often unacknowledged, uses in our everyday lives.

Oil palm

Elaeis guineensis Jacq.; Arecaceae

MODERN WONDER OIL

Many of the plants upon which westerners rely have been important to us for hundreds, if not thousands, of years. An exception is the oil palm. It was only during the twentieth century, as our lives were industrialized, that palm oil became ubiquitous in products as diverse as foods and soaps through industrial lubricants to engine fuels. In 2021 more than 80 million tonnes of palm oil were produced, accounting for about 42 per cent of the global edible oil market. With increasing global interest in biodiesel as a liquid fuel, these numbers will increase.

Crop-growing has transformed landscapes, but usually over hundreds of human generations, in environments where Western societies were unconcerned, indifferent or actively hostile to the consequences of their actions. This is not the case with the oil palm. The drama of its emergence as a significant crop and the transformations its cultivation has wrought on landscapes have happened in just two generations in an environment of global communication where some people care about the fate of the natural world.

Oil palm is indigenous to and widespread in the riverine areas of West and Central Africa. The palms, which can reach 30 metres in height, develop fruit about the size of a plum, with a shiny, reddish brown coat surrounding a thick layer of oil-rich flesh protecting the seed or stone. The oil palm is one of the most economically valuable plants in West Africa, and consequently humans have played a major part in spreading the palm into areas, such as forest clearings, where its natural dispersers, including chimpanzees and palm vultures, would not take it.[456] All parts of the palm have been used for millennia, but oil, extracted from the fruit's flesh, and palm wine, made from fermented sap, are the most significant uses. As might be expected for such an important plant, the peoples of West Africa developed rich vocabularies for both the palm and its products.

As European travellers started to visit West African shores, from the fifteenth century, they saw the ubiquity of the palm in the everyday lives of indigenous peoples, and even reported it in the Americas, where Africans had been enslaved; presumably these palms had been imported by slavers, slave owners or the slaves themselves. The palm was introduced to British glasshouses as early as 1730, and Java's Bogor Botanic Garden received four palms from Mauritius in 1848,[457] but to most Western Europeans they were little more than curiosities to adorn botanical collections at home or in their colonies.

As the nineteenth century drew to a close, some British and North American businessmen realized palm oil could be more than just a product suitable for West African domestic economies; it had global value. Entrepreneurs started to acquire large tracts of West Africa and clear them for oil palm plantations – the palm oil rush had started. Palm oil is a saturated vegetable fat rich in palmitic acid; its chemical composition means it has multiple industrial uses, especially as a lubricant and raw material for soap manufacture. Products from the early years of the industrial exploitation of oil palm, such as Palmolive and Sunlight soaps, became iconic, global brands.

Regions outside Africa, particularly in South East Asia, saw the economic potential of palm oil and began to experiment with their own plantations. Seeds from the four oil palms growing in the Bogor Botanic Garden were distributed across South East Asia, and they became the basis of the commercial plantations started in Sumatra (1911) and Malaysia (1917).[458] The palms established at Deli, in northern Sumatra, produced particularly large quantities of high-quality oil, and by 1930 the so-called 'Deli' palms were being used to established plantations across South East Asia.

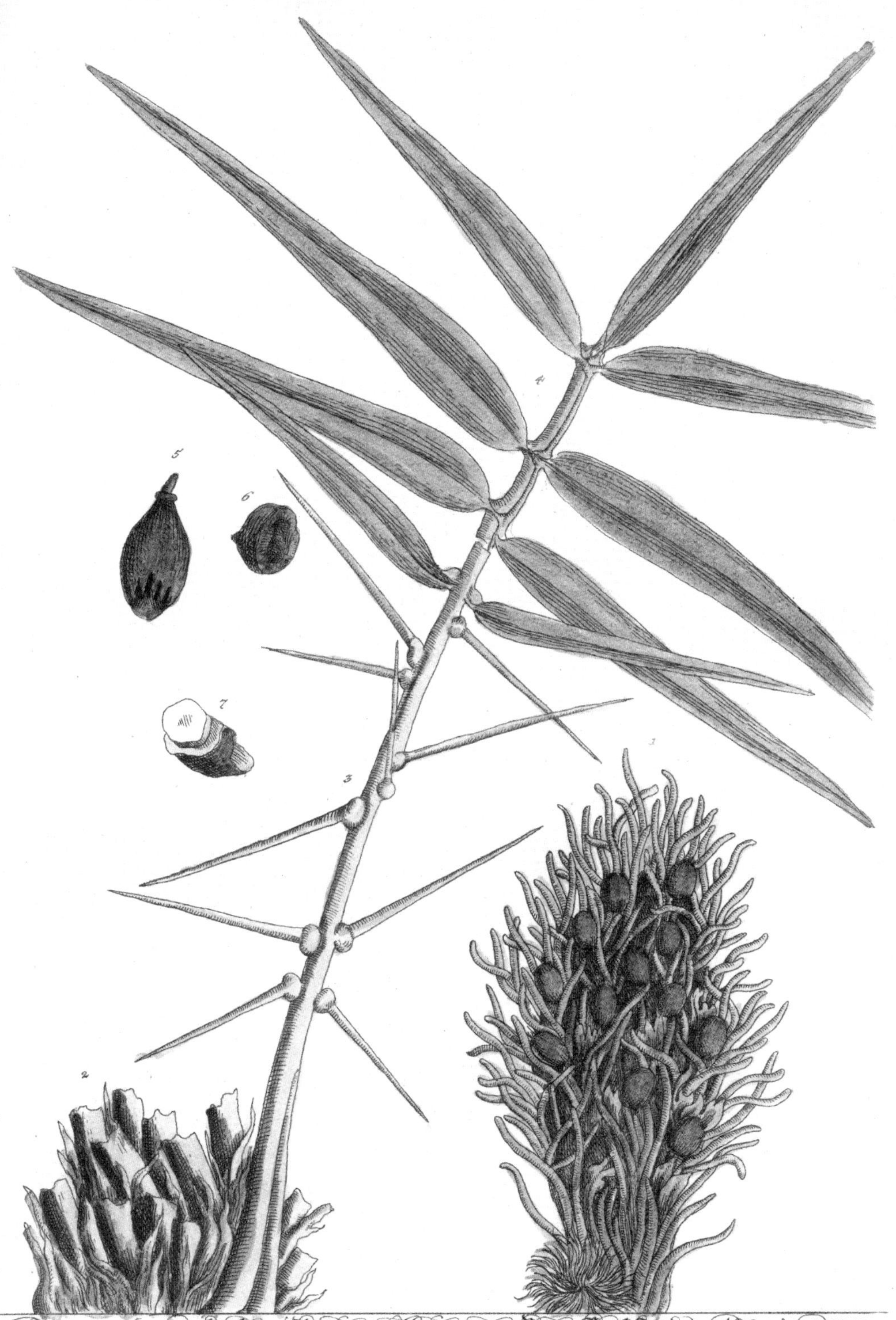

Palma oleosa.

1. der Gipfel des Baums verkleinert, von Herrn Hanß Sloan.
2. ein theil von dem Stam.
3. unterer, 4. oberer Theil eines grünen Blats.
5. die Frücht, 6. offene Schahle, 7. der Kern mit einem theil der Schahle.

Oehl-Palm Baum.

Natural variations in West African oil palms conveniently divide into three types, based on the form of the stone inside the fruit. The most common is the *dura* type, with thick-walled stones. The stones of the other two, less common, types have either thin walls (*tenera* type) or no walls (*pisifera* type). Although the *pisifera* type was not commercially viable, it proved a very important source of breeding material, since when crossed with the *dura* type it produces the highly productive *tenera* type. Plantations in South East Asia were established from tall, particularly productive, *dura*-type individuals introduced to Java. However, formal oil palm breeding in Asia has produced short, easily harvested, 'Dumpy' palms.

Oil palm illustrates the rate at which we can subvert a plant's genetics to our own desires and change landscapes in response to political and economic pressures. In little over a century, palm oil has gone from being harvested from wild, at best semi-domesticated, individual trees to being a commercial operation involving highly bred domesticates. Concomitant with this has been the transformation of forested landscapes. Before the Second World War, West Africa was the largest producer and exporter of palm oil, the creation of palm plantations having happened long before Western consumers were interested in the environmental, ethical or social issues of forestry and agriculture. In the post-war period oil palm production increased exponentially: in the early 1960s approximately 13.6 million tonnes of oil palm fruits were harvested a year; by 2022 this had increased to 424 million tonnes. The bulk of this increased production has happened in Indonesia and Malaysia over the last two decades. However, meeting Western demand for palm oil has come at considerable social and environmental costs in the regions of production. Over the past fifty years areas of Malaysian and Indonesian forest totalling approximately the size of Iceland have been converted to oil palm plantation to satisfy our desire for palm oil.

It has never been easier to collect and analyse large volumes of production data and land use information. Furthermore, the global political climate has changed as consumers, politicians and policymakers have become aware of environmental issues. Western consumers can no longer claim ignorance of the consequences that their decisions have on the lives and environments of the palm oil producers. The orang-utan has become the 'poster primate' for one lobby involved in our subtle and complex relationship with palm oil. In his Foreword to the *Global Biodiversity Outlook 3* report in 2010, the secretary-general of the United Nations, Ban Ki-moon, warned that 'the principal pressures leading to biodiversity loss are not just constant but are, in some cases, intensifying.'[459] But even the doleful gaze of the charismatic orang-utan appears to have failed to slow the felling of forest for oil palm plantations in South East Asia. In 2018 one international conservation organization accepted that 'there appears to be no straightforward way to phase out palm oil without incurring potentially more significant environmental and social impacts elsewhere.'[460]

Some people are optimistic about palm oil in the twenty-first century because of the progress palm breeders have made in transforming strains that were virtually undomesticated. They started their work in the age of genetics, and have been able to take full advantage of plant-breeding technologies. Detailed sequence information about the genetic make-up of a plant is today seen as essential in plant breeding. Publication of the 1.8 billion 'letters' that make up the oil palm genetic code has led some breeders to claim they will be able to contribute to balancing the concerns of industrial and environmental lobbies and produce more palm oil, less destructively.[461] Whether there is a foundation to this optimism remains to be seen, but it is unlikely that habitats already lost will be restored. Yet, we must be careful about criticizing others over how they have changed their environments to benefit their crops, when Western cultures have transformed their own, as well as the environments of others, for millennia.

Calijx
Flosculi facies post:
ant.
later.
大豆
ダイヅ

Soya

Glycine max (L.) Merr.; Fabaceae

SUCCESS STORY OR ECO-DISASTER?

In 1765 a former employee of the British East India Company, Samuel Bowen, brought soya beans from China to Georgia in colonial America. He established the bean as a viable agricultural crop and even exported soya-based foods to Britain.[462] But the production and trade were little more than cottage industries. It was not until more than 150 years later, following the First World War, that farmers, policymakers and industrialists in the USA started to consider soya bean cultivation as a cost-effective way to improve the quality of agricultural soils and provide industrial raw materials. Like other legumes, soya beans have root nodules capable of converting nitrogen in the air into forms of nitrogen that can be taken up by plant roots and hence increase soil fertility (see PEA, page 65). It was soon discovered, especially by researchers working for industrialists such as Henry Ford, that soya beans contain large amounts of edible oil that can be used in everything from food and soap to paint and plastics.[463] Today, soya products reach into every part of our lives. Global soya-based agribusinesses have emerged which dominate international commodity markets and the economies of the regions where soya beans are grown.

Like a number of plants included in the latter part of this book, soya has a quite different history in Eastern and Western cultures. The German explorer and polymath Engelbert Kaempfer illustrated the plant in his *Amoenitatum exoticarum* (1712), describing the uses of soya bean he had observed in Japan,[464] and living plants had been raised in some European botanic gardens since the early eighteenth century, but soya was still little more than a botanical rarity when Carolus Linnaeus christened it in 1759 with the first of its many scientific names. But the lateness of soya's emergence into the Western world belies its ancient roots in Eastern cultures. Domesticated soya beans appear to have been selected from wild soybean, probably in north-east China, with evidence that domesticated plants were cultivated in Korea as early as the tenth century BCE.[465] In the East, beans, liquidized, powdered and turned into curd, have long been used as protein- and oil-rich human food, and the young shoots as vegetables. In contrast, in the West soya beans are mainly used as source of cheap edible oils and protein-rich meal for animal feed manufacture.

Soya beans demonstrate the complex relationship between free trade, the choices consumers make and the natural environment. Global meat consumption is increasing, as populations, especially in Asia, become more affluent.[466] Meeting demand requires cheap protein-rich animal feed. Modern industrial animal farming, and the cheap meat and dairy products it produces, would be impossible without feeds made from soya beans. In 1970 global soya bean production was less than 50 million tonnes and the USA supplied more than 70 per cent of the planet's needs. In 2022 global soya bean production was more than 348 million tonnes, with the United States having about 33 per cent of the market share. The top producer was Brazil with about 35 per cent of the market, whilst Argentina took third place with about 13 per cent of the market.

In the late 1990s soya beans became part of the argument over the acceptance of genetically modified organisms, especially in Europe. In 1995 the American agrochemical company Monsanto introduced soya beans that had been genetically modified to be tolerant to Monsanto's glyphosate herbicide. This was one of the first commercial, genetically modified (GM) crops released to farmers for wholesale planting.

With remarkable naivety, Monsanto assumed that soya bean consumers worldwide would accept GM crops into their diets. They soon discovered European consumers were resistant to such impositions, and a major cultural and trade split opened up between Europe and North America. The GM arguments, with their claims and counter-claims, are complex, involving consumers, commercial companies, governmental regulators, non-governmental organizations, farmers, scientists and billions of pounds in revenue. Arguments abound over labelling, the role of regulators, the independence of research results, the environmental and human health effects of GM crops, the impact of GM crops on farmers, and whether GM crops can even feed the growing world population. In 1997 about 8 per cent of all soya beans cultivated for the commercial market in the USA were GM; by 2023 this had risen to 95 per cent.[467] The global trade in soya beans is now so extensive that GM arguments have been rendered *hors de combat*.

Over the past fifty years, global soya bean yields – that is, the amount of crop per hectare – have approximately doubled. Soya beans require huge areas of flat, well-watered land in an equable climate. Until the 1980s land demands were satisfied by converting North American prairies to soya production, but then commercial interests identified the vast savanna in the centre of Brazil, the *cerrado*, as having agricultural potential. The *cerrado* occupies about a quarter of Brazil's territory, contains about a third of her plant and animal species and is internationally recognized as one of thirty-six global biodiversity hotspots.[468] Agricultural exploitation of the *cerrado* had been limited by the region's acidic, aluminium-rich, nutrient-poor soils, until an American agronomist, Colin McClung, discovered in the 1950s that these impediments could be eliminated by treating *cerrado* soil with lime, gypsum and phosphate fertilizer. Aluminium and acidity levels fell as calcium, magnesium and other nutrient levels increased. With appropriate public and private investment and incentives the stage was set for the *cerrado* region to be stripped of its native vegetation, the rusty-red iron-rich soils turned white with annual fertilizer application and the land transformed to a twenty-first-century South American grain basket. In 2006 McClung and two Brazilian colleagues, Edson

Lobato and Alysson Paolinelli, shared the World Food Prize for their work in transforming agriculture in the *cerrado*.

Today, the *cerrado* is the Brazilian agricultural frontier, where the hides of powerful commercial and political interests are only occasionally penetrated by those with environmental and indigenous-rights concerns. Change in the *cerrado* has been both impressive and alarming. The socio-economic position of people living in the *cerrado* region has been transformed. In a generation the area of Brazil devoted to soya bean cultivation has tripled, and today more than 65 million tonnes of soya bean are produced in an area approximately that of the United Kingdom. Most of the land has come at the expense of the *cerrado*. Estimates suggest that by 2002, 47 per cent of the *cerrado* region had been lost to human development.[469] In 2010 the rate of *cerrado* loss was about 20,000 km² per year, which in 2019 had fallen to about 6,000 km² per year. However, during the 2020s *cerrado* loss was again on the rise (in 2023 approximately 11,000 km² was lost).

The plight of the *cerrado*, described as Brazil's 'forgotten jewel', has been eclipsed by the success of organizations campaigning for the Amazonian rainforest. Advocates of agricultural expansion in the region emphasize the distance of developments from Amazonia rather than their proximity to the *cerrado*; in comparison with the lush richness associated with rainforest, the *cerrado* is seen as little better than worthless scrub.[471]

The prospect that the technology of *cerrado* transformation will move into savanna regions in northern South America and central and southern Africa has been hailed as bringing the benefits of high-yield agriculture to some of the world's poorest places. However, it also brings the prospect of ecological blight, further eroding species diversity and the world's wild places. Yet, if we are to feed large numbers of people aspiring to Western living standards, we have to accept that this cannot be done without environmental damage. Where laws remain weak and the commercial pressure to clear land is high, land will continue to be 'lost' to agriculture, whatever long-term biodiversity arguments might be produced. It remains to be seen whether schemes and dreams of ecological sustainability to remedy past actions are too little, too late. We may even need to question whether the agricultural models used over the last century are appropriate to feed the 9.7 billion people the planet is expected to support in approximately twenty-five years' time.

Corncockle

Agrostemma githago L.; Caryophyllaceae

A CONSERVATION WEATHERGLASS

Familiar and commonplace species may be used as barometers of change. Over millennia we have tamed and domesticated landscapes to grow food and create living space for ourselves. Some plants adapted to these new conditions, and thrived; others moved away, or their ranges contracted, or they became extinct. Such changes were of little concern to agriculturalists thousands of years ago unless they affected their crops, but today they raise concerns over how we affect our environments. Adaptation and extinction are natural biological processes, but the rates at which these processes have occurred over the past century show that we are living through a period of mass species extinction for which we are responsible.[472]

Corncockle is a distinctive annual plant that has been associated with European agricultural landscapes for at least 4,000 years.[473] For much of this period harvested grain was probably contaminated with corncockle seeds, causing problems for millers and consumers, and then sown inadvertently with any seed corn saved from the harvest.[474] Consequently, corncockle often became a troublesome weed. It was common in sixteenth-century Europe, with the herbalist John Gerard stating that 'what hurt it doth among corne, the spoyle unto bread, as well in colour, taste, and unwholesomnes, is better known than desired.'[475] However, some commentators were rather more mindful of it as 'an ornament of our fields'[476] – corncockles, softly hairy and growing to waist height, produce attractive large pink-purple flowers followed by large seed capsules.

a. Lychnis segetum seu Ni= gellastrum, Niellé de Bleds, Raden. b. Lychnis segetum meridiona= lium annua hirsuta floribus al= bis uno verso di= spositis. c. Lychnis saponaria dicta, Saponaire, Seiffen= Kraut. d. Lychnis Saponaria dicta flore multiplici, Herbe à Foulon, Welsch Kraut.

Similar views – a pretty plant, but a nuisance – were being expressed in the early nineteenth century, but by the end of the century people had started to notice the corncockle was disappearing.[477] By the end of twentieth century this was a widespread European trend.

Arable weeds in general are not species one would typically imagine as conservation icons. As formerly common species that probably evolved with us during our 10,000-year-long adventure with agriculture, they might be expected to acclimate readily to environmental change. Factors responsible for arable weed disappearance are associated with the mechanization of agriculture, improvements in seed-cleaning processes and the use of herbicides and fertilizers.[478] For the farmer, concerned with food production and profit, the disappearance of arable weeds is a boon. For the environmentalist, their decline is symptomatic of wider environmental problems. However, despite the concern over corncockle in Europe, the species does not face global extinction. It remains a significant pest of agriculture in other parts of the world, where European concerns over its decline must seem like an indulgence of the well fed.

In 1864 American diplomat George Marsh created a vivid precautionary image when appealing for his fellow Americans to concern themselves with the future of their environment:

> we are, even now, breaking up the floor and wainscoating and doors and window frames of our dwelling, for fuel to warm our bodies and seethe our pottage, and the world cannot afford to wait till the slow and sure progress of exact science has taught it a better economy.[479]

Although Marsh's concerns were unorthodox for the period, they were not unique,[480] and are now voiced with greater urgency. The past forty years have seen an increased awareness, if not acceptance, of the environmental issues associated with species conservation in the face of human activities. Central to our effects on the environment is our relentlessly expanding population, which means we destroy, pollute and overexploit increasing areas of the natural environment to maintain or improve our lifestyles.

During the late 1980s, 'biodiversity', the variety of life on Earth, became a buzzword liberally flung around by politicians, business leaders, media commentators, scientists and the public alike, as public and private interests jockeyed for position in the emerging environmental debate. At the heart of the concept of biodiversity are the number and variability of species, often the backbone of conservation agendas. Plant conservation agendas built on potential, often unknown, medicinal and agricultural values of plants have shifted to embrace the idea that our environment provides essential services for our well-being.

By 1992 the central role of non-human life in human social and economic development was recognized in the Convention on Biological Diversity, which focused on national strategies for the conservation and sustainable use of biological diversity.[481] To date, the vast majority of countries have signed and ratified the Convention; the biggest omission is the USA. In 2002 a sixteen-point Global Strategy for Plant Conservation was adopted internationally to halt the loss of plant diversity by 2020.

The biologically implausible tale of Noah and his Ark can be portrayed as an environmental story, although Noah shows a complete indifference to the central role of plants in his life and the lives of the animals he tried to save. Nevertheless, the Ark has become a powerful conservation metaphor, especially when applied to the world's great seed banks, such as the Svalbard Seed Vault and Kew's Millennium Seed Bank at Wakehurst. Such programmes illustrate the grand scales upon which conservation programmes can be conceived and implemented.[482] However, conservation of arable weeds, such as corncockle, requires much smaller, albeit heavily subsidized, actions by farmers.[483]

One hundred and fifty years on from George Marsh's prescient warning, the science of the environment has progressed in ways inconceivable to him. We know more than ever about the environmental threats we face and the scales at which we threaten the environment; environmental concerns have become part of everyday language, even if only to naysay the environmental agenda. However, conservation remains dictated by the twin pillars of politics and money: the politics of short-term image and the dream of financial exploitation of the environment at no cost. In Western cultures we are starting to (re)recognize we are tenants and environmental stewards, which some argue non-Western cultures never lost. We must take responsibility for our own future rather than evoke fate and divine judgement or dreams of distant stars and extraterrestrial human colonies.

Thale cress

Arabidopsis thaliana (L.) Heynh.; Brassicaceae

A MODEL PLANT

At first sight, this final plant is useless. We began with barley, with its vital contributions to food and drink, and its historical role in trade; we end with a plant that has neither economic nor cultural value. It is not edible or medicinal; it does not even have attractive flowers. Thale cress is not mentioned in the myths and legends of classical Western civilizations. It escapes the notice of archaeologists. It has not inspired great art. Over the last two decades, however, research that started in thale cress has forced governments, non-governmental organizations, multinational companies and individuals to confront the ethical, philosophical, economic and political issues of pure and applied botanical research that bind cultures together.

Thale cress has probably been around in the temperate regions of the Old World northern hemisphere for at least as long as we have been disturbing and cultivating the land. It is a rapidly growing little plant that produces a rosette of small, hairy, spoon-shaped leaves. The tiny white flowers, hanging on a stem that emerges from the centre of the rosette, eventually produce long, narrow fruits. The whole plant rarely gets more than 25 cm tall. The first identifiable mention of this diminutive plant was in 1588, in an obscure account of the Harz forest flora by the Thuringian physician Johannes Thal, published posthumously.[484] Tucked away among the woodcuts is an illustration of a plant Thal called *Pilosella siliquata*. In 1753 Carolus Linnaeus used Thal's illustration when he described *Arabis thaliana*. The weedy little plant does have recognizable similarities to the more colourful Arabis species popular as garden rock plants, but about a century later it was

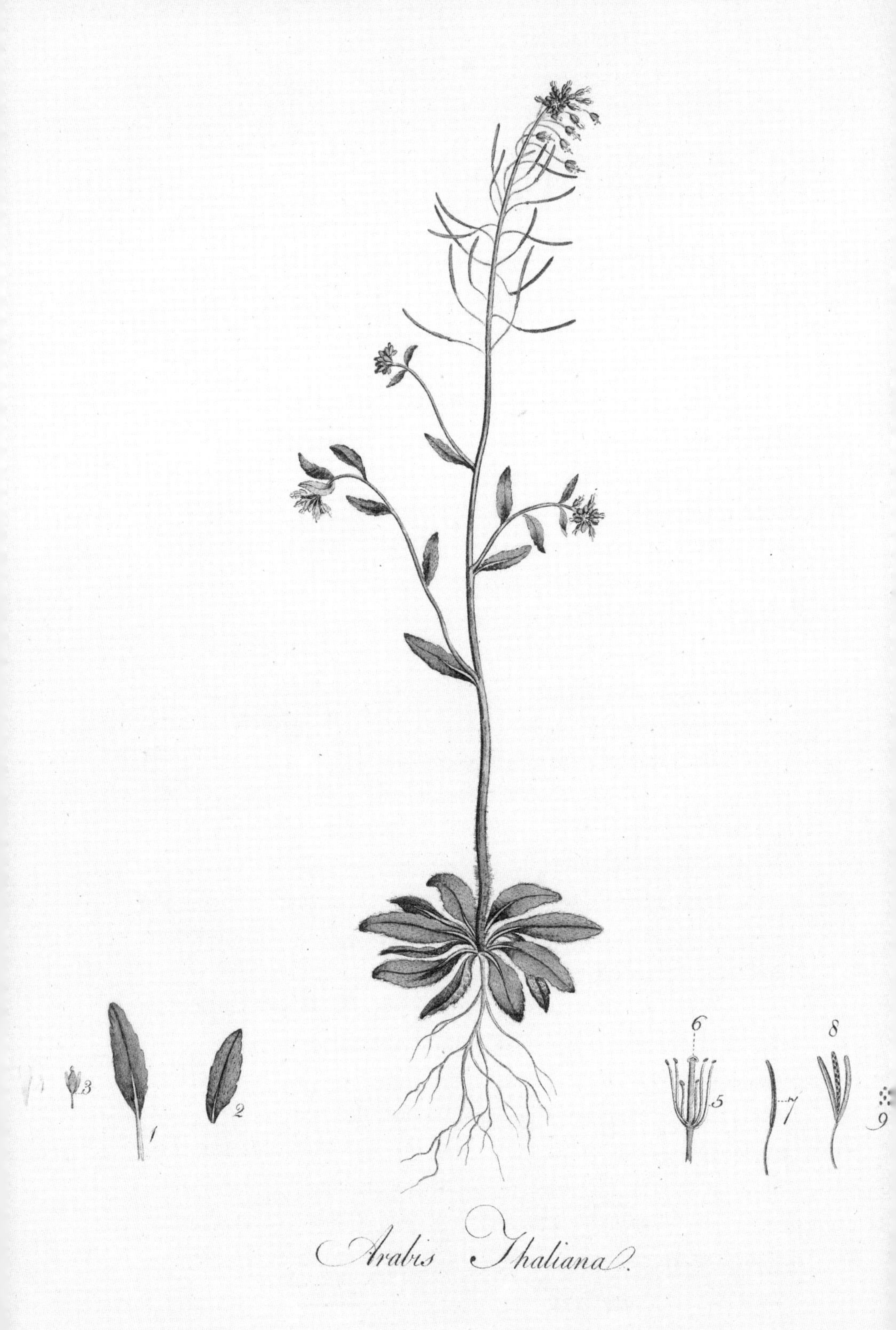

Arabis Thaliana.

considered sufficiently different to be placed in its own genus, *Arabidopsis* (meaning '*Arabis*-like').[485]

During the eighteenth and nineteenth centuries botany was synonymous with the description of plants' diversity, morphology and anatomy. By the nineteenth century scientists were realizing that description was not enough; to understand how plants functioned in detail, it was necessary to experiment. Charles Darwin and Alfred Russel Wallace were promulgating their thought-changing ideas on evolution, variation and adaptation, while Gregor Mendel showed the value of experimental approaches to understanding how features were inherited from one generation to the next (see PEA, page 65). Other researchers used experimentation to study all aspects of plants' lives, from population and evolutionary biology through physiology and biochemistry to cell and developmental biology. Experimentation allowed ideas to be tested, explaining the patterns of diversity recorded; experimental and descriptive approaches to plant biology complemented each other.

By the early twentieth century, when the German botanist Friedrich Laibach was searching for a model species to investigate plant chromosomes, the experimental approach was firmly established. For some reason, Laibach chose thale cress, but abandoned it when he discovered it had tiny chromosomes. However, Laibach and his students eventually returned to thale cress to investigate the mutagenic effects of radiation and built a collection of mutants. After the Second World War, Laibach continued to build his collection so that it encompassed thale cress ecotypes from across the globe. By the mid-1960s the case for using thale cress as a model plant was strong, and by the late 1980s, with the backing of national and international funding agencies, the global scientific community embraced it wholeheartedly. Microbiologists and geneticists had adopted phages and fruit flies as their model organisms; plant scientists now had their own botanical workhorse.

Thale cress has proven to be an excellent model plant.[486] It has one of the smallest nuclear DNA sequences (genomes) known for any plant, comprising some 157 million base pairs divided across five chromosomes. (For comparison, the human genome is about 3.3 billion base pairs long divided across 23 chromosomes.[487]) When the complete thale cress genome sequence was published in the year 2000, it was the first plant to have its DNA fully decoded.[488] Over the last decade, functions have been assigned to more than 25,000 of the encoded genes and proteins.[489] For genetic research, the plant's small physical size makes it convenient for growing in vast numbers, while the short life cycle (approximately six weeks from germination to seed production) means many generations can be produced in a single year. Furthermore, thale cress regularly sets large quantities of selfed seed, so genetic lines can be maintained, but it is also readily crossed. Thale cress can be routinely transformed to create genetically modified experimental plants in which to explore gene function and localization. Importantly, data and genetic information are shared among research groups, while seed and DNA stocks are readily available through international resource centres.

Over the last two decades we have acquired vast amounts of knowledge about the fundamental working of plants using thale cress. Most of this knowledge has come through pure research, conducted because particular researchers were curious rather than with an applied goal in mind. Such pure research (perhaps strategic research would be a better phrase), funded by governments, charities and private companies, is criticized by some as having no value. However, such criticisms fail to recognize that research outcomes are unpredictable, and that we never know what might be useful in the future. For example, using thale cress it was possible to test a fundamental hypothesis about how three genes interact to produce the typical parts (sepals, petals, stamens and carpels) of all flowers on the planet.[490] The German botanist Alexander Braun produced the first description of a thale cress mutant, a double-flowered type in 1873; it was not until 1990 that a genetic explanation of this mutant was published.[491] Our knowledge of thale cress biology is extensive but it does not explain all of the plant diversity we see around us; the more we have learnt, the more we have come to realize that thale cress has its limitations and more model plants are needed.

As our pure knowledge of thale cress genes has increased, so practical applications of this knowledge have been investigated. Genes whose functions were tested in thale cress can be used to give them characteristics that they may not naturally acquire. Such organisms have been the source of considerable, often polarized, discussion. The spectres of toxic food and environmental devastation on one side of the debate are contrasted with solutions to feeding and healing the growing human population on the other.

Ten thousand years ago, plant domestication was a happy accident. Today, our unparalleled knowledge of the genetics of thale cress, and other species, presents us with challenges about whether we should do things just because we can. These issues are not solely the remit of scientists or specific lobby groups but, in a hopelessly idealistic world, are challenges to us all, since they are at the heart of our cultures.

Notes

1. Townsend, T. (1738) *The history of the conquest of Mexico by the Spaniards. Translated into English from the original Spanish of Don Antonio de Solis*, Vol. 1. Printed for John Osborn, London, p. 416.
2. Pironon, S., Ondo, I., Diazgranados, M., Allkin, R., Baquero, A.C., Cámara-Leret, R., Canteiro, C., Dennehy-Carr, Z., Govaerts, R., Hargreaves, S., Hudson, A.J., Lemmens, R., Milliken, W., Nesbitt, M., Patmore, K., Schmelzer, G., Turner, R.M., van Andel, T.R., Ulian, T., Antonelli, A., and Willis, K.J. (2024) The global distribution of plants used by humans. *Science* 383: 293–7.
3. Ebbell, B. (1937) *The Papyrus Ebers: The greatest Egyptian medical document.* Oxford University Press, London. Thompson, R.C. (1949) *The Assyrian Herbal.* Luzac, London.
4. Gunther, R.T. (1934) *The Greek Herbal of Dioscorides. Illustrated by a Byzantine, A.D. 512; Englished by John Goodyer, A.D. 1655.* University Press, Oxford. Arber, A. (1986) *Herbals.* Cambridge University Press, Cambridge.
5. De Candolle, A. (1884) *Origin of cultivated plants.* Kegan Paul, Trench & Co., London.
6. Harlan, J.R. (1975) *Crops & man.* American Society of Agronomy, Madison WI.
7. Smith, B.D. (1998) *The emergence of agriculture.* Scientific American Library, New York.
8. Marsh, G.P. (1864) *Man and nature; or, Physical geography as modified by human action.* Charles Scribner, New York, p. 36.
9. Rackham, O. (1995). *Trees and woodlands in the British landscape: The complete history of Britain's trees, woods and hedgerows.* Weidenfeld & Nicolson, London.
10. Fowler, W.W. (1908) *The Roman festivals of the period of the Republic.* Macmillan, London.
11. Dai, F., Nevo, E., Wu, D., Comadran, J., Zhou, M., Qiu, L., Chen, Z., Beiles, A., Chen, G., and Zhang, G. (2012). Tibet is one of the centers of domestication of cultivated barley. *Proceedings of the National Academy of Sciences of the United States of America* 109: 16969–73.
12. Michel, R.H., McGovern, P.E., and Badler V.R. (1993) The first wine and beer: Chemical detection of ancient fermented beverages. *Analytical Chemistry* 65: 408A–413A.
13. George, A.R. (1999) *The Epic of Gilgamesh.* Penguin Books, London.
14. Manchester K.L. (1995) Louis Pasteur (1822–1895) – chance and the prepared mind. *Trends in Biotechnology* 13: 511–15.
15. Dillon, P. (2002) *The much-lamented death of Madam Geneva.* Review, London.

Valverde, M. (1998) *Diseases of the will: Alcohol and the dilemmas of freedom.* Cambridge University Press, Cambridge.
16. Ridgeway, W. (1892) *The origin of metallic currency and weight standards.* Cambridge University Press, Cambridge, pp. 180–81.
17. Holloway, S.W.F. (1991) *Royal Pharmaceutical Society of Great Britain 1841–1991: A political and social history.* Pharmaceutical Press, London.
18. Jones, W.H.S. (2001) *Pliny: Natural history. Books 24–27.* Harvard University Press, London, p. 273.
19. Wink, M., and van Wyk, B.-E. (2008) *Mind-altering and poisonous plants of the world.* Timber Press, Portland OR.
20. Lewis, W.H., and Elvin-Lewis, M.P.F. (2003) *Medical botany: Plants affecting human health.* John Wiley, Hoboken NJ.
21. Rudgley, R. (1998). *The encyclopaedia of psychoactive substances.* Abacus, London.
22. Bennett, C.E. (1925) *Frontinus: The Stratagems and The Aqueducts of Rome.* Heinemann, London, ch. 2, 5.
23. Shakespeare, W., *Othello*, III.iii.
24. Fleisher, A., and Fleisher, Z. (1994) The fragrance of biblical mandrake. *Economic Botany* 48: 243–51.
25. *Genesis* 30:14–16.
26. Harner, M.J. (1974) The role of hallucinogenic plants in European witchcraft. In Harner, M.J. (ed.) *Hallucinogens and shamanism.* Oxford University Press, Oxford, pp. 125–50.
27. Rudgley R. (1998). *The encyclopaedia of psychoactive substances.* Abacus, London.
28. Shakespeare, W., *Romeo and Juliet*, IV.iii.
29. Gerard, J. (1633) *The herball or Generall historie of plantes. Gathered by Iohn Gerarde of London Master in Chirurgerie very much enlarged and amended by Thomas Iohnson citizen and apothecarye of London.* Printed by Adam Islip Ioice Norton and Richard Whitakers, London, p. 351.
30. Gerard, J. (1633) *The herball or Generall historie of plantes. Gathered by Iohn Gerarde of London Master in Chirurgerie very much enlarged and amended by Thomas Iohnson citizen and apothecarye of London.* Printed by Adam Islip Ioice Norton and Richard Whitakers, London, p. 351.
31. Grover, N. (1965) Man and plants against pain. *Economic Botany* 19: 99–112.
32. Rivera, D., Obón, C., Heinrich, M., Inocencio, C., Verde, A., and Farajado, J. (2006) Gathered Mediterranean food plants – ethnobotanical investigators and historical development. In Heinrich, M., Müller, W.E., Galli, C. (eds) *Local Mediterranean food plants and nutraceuticals.* Karger, Basel, pp. 18–74.
33. Magnol, P. (1686) *Botanicum Monspeliense.* Ex Officina Danielis Pech, Monspelii, p. 38.
34. Vilmorin-Andrieux, M. (1885) *The vegetable garden.* John Murray, London.
35. Marggraf, A. (1747) Expériences chimiques faites dans le dessein de tirer un véritable sucre de diverses plantes, qui croissent dans nos contrées. *Histoire de l'académie royale des sciences et belles-lettres de Berlin* 3: 79–90.
36. Fischer, H.E. (1989) Origin of the 'Weiße Schlesische Rübe' (white Silesian beet) and resynthesis of sugar beet. *Euphytica* 41: 75–80.
37. Dohm, J.C., Minoche, A.E., Holtgrawe, D., Capella-Gutierrez, S., Zakrzewski, F., Tafer, H., Rupp, O., Sorensen, T.R., Stracke, R., Reinhardt, R., Goesmann,

A., Kraft, T., Schulz, B., Stadler, P.F., Schmidt, T., Gabaldon, T., Lehrach, H., Weisshaar, B., and Himmelbauer, H. (2014) The genome of the recently domesticated crop plant sugar beet (*Beta vulgaris*). *Nature* 505: 546–9.

38. Griggs, B. (1997) *Green pharmacy: The history and evolution of western herbal medicine*. Healing Arts Press, Rochester VT.
39. Sala, A. (1618) *Opiologia or, a treatise concerning the nature, properties, true preparation and safe vse and administration of opium. And done into English, and something inlarged by Tho. Bretnor*. Printed by Nicholas Okes, London.
40. Latham, R.G. (1850) *The works of Thomas Sydenham, M.D.*, Vol. 2. Sydenham Society, London, p. 98.
41. Poe, E.A. (1845) in Poe, E.A., *Tales*. Wiley & Putnam, London, in his short story 'Ligeia' (1838). Kramer, J.C. (1979) Opium rampant: Medical use, misuse and abuse in Britain and the west in the 17th and 18th centuries. *British Journal of Addiction* 74: 377–89.
42. Morimoto, S., Suemori, K., Moriwaki, J., Taura, F., Tanaka, H., Aso, M., Tanaka, M., Suemune, H., Shimohigashi, Y., Shoyama, Y. (2001) Morphine metabolism in the opium poppy and its possible physiological function: Biochemical characterization of the morphine metabolite, bismorphine. *Journal of Biological Chemistry* 276: 38179–84.
43. Schriff, P.L. (2002) Opium and its alkaloids. *American Journal of Pharmaceutical Education* 66: 168–94. Askitopoulou, H., Ramoutsaki, I.A., and Konsolaki, E. (2002) Archaeological evidence on the use of opium in the Minoan world. *International Congress Series* 1242: 23–9.
44. Holloway, S.W.F. (1991) *Royal Pharmaceutical Society of Great Britain 1841–1991: A political and social history*. Pharmaceutical Press, London.
45. Kramer, J.C. (1979) Opium rampant: Medical use, misuse and abuse in Britain and the west in the 17th and 18th centuries. *British Journal of Addiction* 74: 377–89.
46. Zheng, Y. (2003) The social life of opium in China, 1483–1999. *Modern Asian Studies* 37: 1–39. Brown, R.H. (2002) The opium trade and opium policies in India, China, Britain, and the United States: Historical comparisons and theoretical interpretations. *Asian Journal of Social Science* 30: 623–56.
47. Fairbanks, J.K. (1992) The creation of the treaty system. In Fairbanks, J.K. (ed.) *The Cambridge History of China*, Vol. 10, Part 1. Cambridge University Press, Cambridge, p. 213.
48. Dikötter, F., Laamann, L., and Xun, Z. (2004) *Narcotic culture: A history of drugs in China*. University of Chicago Press, Chicago.
49. Peters, G. (2009) *Seeds of terror: How heroin is bankrolling the Taliban and Al Qaeda*. Hachette India, Gurgaon, India. UN Office of Drugs and Crime (2023) Press Release: Myanmar overtakes Afghanistan as world's top opium producer. https://news.un.org/en/story/2023/12/1144702 (accessed 25 December 2023).
50. Bloch-Dano, E. (2013) *Vegetables: A biography*. University of Chicago Press, Chicago.
51. Morison, R. (1680) *Plantarum Historiae Universalis Oxoniensis pars secunda*. Threatro Sheldoniano, Oxonii, pp. 208–9.
52. Hedrick, U.P. (1919) *Sturtevant's notes on edible plants*. J.B. Lyon Company, Albany, p. 211.

53. Darwin, C.R. (reprint 1998) *The variation of animals and plants under domestication*, Vol. 1. Johns Hopkins University Press, Baltimore MD, p. 322.
54. Li, H.L. (1970) The origin of cultivated plants in Southeast Asia. *Economic Botany* 24: 3–19.
55. Song, K.M., Osborn, T.C., and Williams, P.H. (1988) *Brassica* taxonomy based on nuclear RFLPs. I. Genome evolution of diploid and amplidiploid species. *Theoretical and Applied Genetics* 75: 784–94. Palmer, J.D., Shields, C.R., Cohen, D.B., and Orton, T.J. (1983) Chloroplast DNA evolution and the origin of amphidiploid *Brassica* species. *Theoretical and Applied Genetics* 65: 181–9. Allender, C.J., and King, G.J. (2010) Origins of the amphiploid species *Brassica napus* L. investigated by chloroplast and nuclear molecular markers. *BMC Plant Biology* 10: 54.
56. Small, E. (1975) On toadstool soup and legal species of marihuana. *Plant Science Bulletin* 21: 34–9.
57. Jiang, H.-E., Li, X., Zhao, Y.-X., Ferguson, D.K., Hueber, F., Bera, S., Wang, Y.-F., Zhao, L.-C., Liu, C.J., and Li, C.S. (2006) A new insight into *Cannabis sativa* (Cannabaceae) utilization from 2500-year-old Yanghai Tombs, Xinjiang, China. *Journal of Ethnopharmacology* 108: 414–22.
58. Postlethwayt, M. (1766) *The universal dictionary of trade and commerce*. Printed for H. Woodfall, London. Anonymous (1844) *The dictionary of trade, commerce and navigation*. Brittain, London.
59. Van Slageren, M.W. (1994) *Wild wheats: A monograph of Aegilops L., and Amblyopyrum (Jaub. & Spach) Eig (Poaceae)*. Agricultural University, Wageningen.
60. Levy, A.A., and Feldman, M. (2005) Genetic and epigenetic reprogramming of the wheat genome upon allopolyploidisation. *Biological Journal of the Linnean Society* 82: 607–15.
61. Heun, M., Abbo, S., Lev-Yadun, S., and Gopher, A. (2012) A critical review of the protracted domestication model for Near-Eastern founder crops: linear regression, long-distance gene flow, archaeological, and archaeobotanical evidence. *Journal of Experimental Botany* 63: 695–709.
62. Martin, D.L., and Goodman, A.H. (2002) Health conditions before Columbus: Paleopathology of native North Americans. *Western Journal of Medicine* 176: 65–8. Pechenkina, E.A., Benfer, R.A., and Zhijun, W. (2002) Diet and health changes at the end of the Chinese Neolithic: The Yangshao/Longshan Transition in Shaanxi Province. *American Journal of Physical Anthropology* 117: 15–36.
63. Rudney, J.D. (1982) Dental indicators or growth disturbance in a series of ancient Lower Nubian populations: Changes over time. *American Journal of Physical Anthropology* 60: 463–70.
64. Jablonski, N.G., and G. Chaplin, G. (2010) Human skin pigmentation as an adaptation to UV radiation. *Proceedings of the National Academy of Sciences USA* 107: 8962–8.
65. Holden, C., and Mace, R. (1997) Phylogenetic analysis of the evolution of lactose digestion in adults. *Human Biology* 69: 605–28. Keller, A., Graefen, A., Ball, M., Matzas, M., Boisguerin, V., Maixner, F., Leidinger, P., Backes, C., Khairat, R., Forster, M., Stade, B., Franke, A., Mayer, J., Spangler, J., McLaughlin, S., Shah, M., Lee, C., Harkins, T. T., Sartori, A., Moreno-Estrada, A., Henn, B., Sikora, M., Semino, O., Chiaroni, J., Rootsi, S., Myres, N.M., Cabrera, V.M., Underhill,

P.A., Bustamante, C.D., Vigl, E.E., Samadelli, M., Cipollini, G., Haas, J., Katus, H., O'Connor, B.D., Carlson, M.R.J., Meder, B., Blin, N., Meese, E., Pusch, C.M., and Zink, A. (2012) New insights into the Tyrolean Iceman's origin and phenotype as inferred by whole-genome sequencing. *Nature Communications* 3: 698. Ingram, C.J.E., Elamin, M.F., Mulcare, C.A., Weale, M.E., Tarekegn, A., Raga, T.O., Bekele, E., Elamin, F.M., Thomas, M.G., Bradman, N., and Swallow, D.M. (2007) A novel polymorphism associated with lactose tolerance in Africa: Multiple causes for lactase persistence? *Human Genetics* 120: 779–88.

66. Mummert, A., Esche, E., Robinson, J., and Armelagos, G.J. (2011) Stature and robusticity during the agricultural transition: Evidence from the bioarchaeological record. *Economics and Human Biology* 9: 284–301.
67. Scarborough, J. (1982) Beans, Pythagoras, taboos, and ancient dietetics. *The Classical World* 75: 355–8. Meletis, J. (2012) Favism: A brief history from the 'abstain from beans' of Pythagoras to the present. *Archives of Hellenic Medicine* 29: 258–63.
68. Danet, P. (1700) *A complete dictionary of the Greek and Roman antiquities*. John Nicholson, Tho. Newborough, John Bullord, R. Parker and B. Cooke, London.
69. Meletis, J. (2012) Favism. A brief history from the 'abstain from beans' of Pythagoras to the present. *Archives of Hellenic Medicine* 29: 258–63.
70. 'Cum faba florescit, stultorum copia crescit'; de Mery, M.C. (1828) *Histoire générale des proverbes, adages, sentences, apophthegmes*. Delongchamps, Libraire-Éditeur, Paris, p. 337.
71. Hopper, WD (2006) *Marcus Terentius Varro: On agriculture*. Harvard University Press, Cambridge MA, ch. 23, verse 3.
72. Lempriere, J. (1812) *Classical dictionary*. T. Cadell and W. Davies, London.
73. Phillips, H. (1822) *History of cultivated vegetables*, Vol. 1. Henry Colburn & Co., London, p. 71.
74. Anonymous (1791) *Observations on the evidence given before the Committees of the Privy Council and House of Commons in support of the Bill for abolishing the slave trade*. John Stockdale, London.
75. Phillips, H. (1822) *History of cultivated vegetables*, Vol. 1. Henry Colburn & Co., London, p. 70.
76. Gerard, J. (1633) *The herball or Generall historie of plantes. Gathered by Iohn Gerarde of London Master in Chirurgerie very much enlarged and amended by Thomas Iohnson citizen and apothecarye of London*. Printed by Adam Islip Ioice Norton and Richard Whitakers, London.
77. Vilmorin-Andrieux, M. (1885) *The vegetable garden*. John Murray, London, p. 22.
78. Duc, G., Bao, S., Baum, M., Redden, B., Sadiki, M., Suso, M.J., Vishniakova, M., and Zong, X. (2010) Diversity maintenance and use of *Vicia faba* L. genetic resources. *Field Crops Research* 115: 270–78.
79. Tanno, K.-I., and Willcox, G. (2006) The origins of cultivation of *Cicer arietinum* L., and *Vicia faba* L.: Early finds from Tell el-Kerkh, north-west Syria, late 10th millennium b.p. *Vegetation History and Archaeobotany* 15: 197–204. Willcox, G., Fornite, S., and Herveux, L. (2008) Early Holocene cultivation before domestication in northern Syria. *Vegetation History and Archaeobotany* 17: 313–25.
80. Pickering, D. (2002). *Cassell's dictionary of superstitions*. Cassell, London.

81. Coles, W. (1657) *Adam in Eden: or, Natures Paradise. The history of plants, fruits, herbs and flowers.* J. Streater, London, pp. 135–7.
82. Stol, M. (1987) Garlic, onion, leek. *Bulletin of Sumerian Agriculture* 3: 57–80. Bottéro, J. (1987) The Culinary Tablets at Yale. *Journal of the American Oriental Society* 107: 11–19. Täckholm, V., and Drar, M. (1954) *Allium* in ancient Egypt. In Täckholm, V., and Drar, M. (eds) *Flora of Egypt*, Vol. 3. Cairo University Press, Cairo, pp. 93–106. Crawford, D. (1973) Garlic-growing and agricultural specialization in Graeco-Roman Egypt. *Chronique d'Égypte* 48: 350–63.
83. Littlebury, I. (1737) *The History of Herodotus*, Vol. 1. Printed for D. Midwinter, A. Bettesworth and C. Hitch, J. and J. Pemberton, R. Ware, C. Rivington, F. Clay, J. Batley and J. Wood, A. Ward, J. and P. Knapton, T. Longman and R. Hett, London, p. 211.
84. King, C.W. (1882) *Plutarch's morals: Theosophical essays.* George Bell & Sons, London, Book 1, 8.
85. Murray, M.A. (2000) Fruits, vegetables, pulses and condiments. In Nicholson, P.T., and Shaw, I. (eds) *Ancient Egyptian material and technology.* Cambridge University Press, Cambridge, pp. 505–36. De Vartavan, C., Arakelyan, A., and Amorós, V.A. (2010) *Codex of ancient Egyptian remains.* SAIS, London.
86. Parejko, K. (2003) Pliny the Elder's Silphium: First recorded species extinction. *Conservation Biology* 17: 925–7.
87. Gerard, J. (1597) *The Herball or general historie of plantes. Gathered by John Gerarde of London Master in Chirurgerie.* John Norton, London, p. 133.
88. Stearn, W.T. (1978) European species of *Allium* and allied genera of Alliceae: a synonymic enumeration. *Annales Musei Goulandris* 4: 83–198. Mathew, B. (1996) *A review of Allium section Allium.* Royal Botanic Garden Kew, Richmond. Zohary, D., Hopf, M., and Weiss, E. (2013) *Domestication of plants in the Old World.* Oxford University Press, Oxford, pp. 155–7.
89. Kamenetsky, R. (2007) Garlic: botany and horticulture. *Horticultural Reviews* 33: 123–72. For an alternative view of the origin of the cultivated onion, see Yusupov, Z., Deng, T., Volis, S., Khassanov, F., Makhmudjanov, D., Tojibaev, K., and Sun, H. (2021) Phylogenomics of *Allium* section *Cepa* (Amaryllidaceae) provides new insights on domestication of onion. *Plant Diversity* 43: 102–10.
90. Eady, C.C., Kamoi, T., Kato, M., Porter, N.G., Davis, S., Shaw, M., Kamoi, A., and Imai, S. (2008) Silencing onion lachrymatory factor synthase causes a significant change in the sulfur secondary metabolite profile. *Plant Physiology* 147: 2096–106.
91. Food and Agriculture Organization of the United Nations (2023) FAOSTAT http://faostat.fao.org (accessed 26 December 2023).
92. Hooper, W.D. (1934) *Cato and Varro on agriculture.* Harvard University Press, Cambridge MA, ch. 38.
93. Fairclough, H.R. (1916) *Virgil: Eclogues, Georgics, Aeneid I–VI.* Harvard University Press, Cambridge MA, pp. 177–8.
94. Weeden, N.F. (2007) Genetic changes accompanying the domestication of *Pisum sativum*: Is there a common genetic basis to the 'domestication syndrome' for legumes? *Annals of Botany* 100: 1017–25.
95. Vilmorin-Andrieux, M. (1885) *The vegetable garden.* John Murray, London, pp. 385–438.

96. Darwin, C.R. (reprint 1998) *The variation of animals and plants under domestication*, Vol. 1. Johns Hopkins University Press, Baltimore MD, pp. 345–9.
97. Henig, R.M. (2000) *A monk and two peas: The story of Gregor Mendel and the discovery of genetics*. Phoenix, London.
98. Knight, T.A. (1799) An account of some experiments on the fecundation of vegetables. In a letter from Thomas Andrew Knight, Esq. to the Right Hon. Sir Joseph Banks, K.B. P.R.S. *Philosophical Transactions of the Royal Society of London* 89: 195–204.
99. MS Sherard 219, fols 142–143 (Sherardian Library of Plant Taxonomy, University of Oxford).
100. MS Sherard 219, fol. 143 (Sherardian Library of Plant Taxonomy, University of Oxford). Harris, S.A. (2021) Sibthorp's *Flora Graeca* expedition and teaching Linnaean botany in Oxford Physic Garden. *Curtis's Botanical Magazine* 38: 451–71.
101. Zohary, D. (1995) Olive, *Olea europaea* (Oleaceae). In Smartt, J., and Simmonds, N.W. *Evolution of crop plants*. Longman, London, pp. 379–82.
102. *Deuteronomy* 8:8.
103. De Vartavan, C., Arakelyan, A., and Amorós, V.A. (2010) *Codex of ancient Egyptian plant remains*. SAIS Academic Books, London.
104. Dunmire, W.M. (2004) *Gardens of New Spain: How Mediterranean plants and foods changed America*. University of Texas Press, Austin TX.
105. Spennemann, D.H.R., and Allen, L.R. (2000) Feral olives (*Olea europaea*) as future woody weeds in Australia: A review. *Australian Journal of Experimental Agriculture* 40: 889–901.
106. Green, P.S. (2002) A revision of *Olea* L. (Oleaceae). *Kew Bulletin* 57: 91–140.
107. Lumaret, R., and Ouazzani, N. (2001) Ancient wild olives in Mediterranean forests. *Nature* 413: 700. Breton, C., Tersac, M., and Berville, A.A. (2006) Genetic diversity and gene flow between the wild olive (oleaster, *Olea europaea* L.) and the olive: Several Plio-Pleistocene refuge zones in the Mediterranean Basin suggested by simple sequence repeats analysis. *Journal of Biogeography* 33, 1916–28.
108. Myles, S., Boyko, A.R., Owens, C.L., Brown, P.J., Grassi, F., Aradhya, M.K., Prins, B., Reynolds, A., Chia, J.-M., Ware, D., Bustamante, C.D., Buckler, E.S. (2011) Genetic structure and domestication history of the grape. *Proceedings of the National Academy of Sciences USA* 108: 3530–35. McGovern, P.E. (2003) *Ancient wine: The search for the origins of viniculture*. Princeton University Press, Princeton NJ.
109. *Genesis* 9: 20–21. Statista (2023) www.statista.com (accessed 26 December 2023).
110. Harvey J.H. (1981) *Mediaeval gardens*. Batsford, London.
111. McGovern, P.E. (2003) *Ancient wine: The search for the origins of viniculture*. Princeton University Press, Princeton NJ.
112. Royer, C. (1988) Mouvement historiques de la vigne dans le monde. In La Manufacture et la Cité des sciences et de l'industrie (ed.) *La Vigne et le Vin*. Graficas, Paris, pp. 15–25.
113. Martin, G. (2003) *Domesday Book: A complete translation*. Penguin, London.
114. Fagan, B. (2000) *The little ice age: How climate made history 1300–1850*. Basic Books, London.
115. Kunz, K., and Sigurdsson, G. (2008) *The Vinland sagas*. Penguin Books, London.

116. Moerman, D.E. (1998) *Native American ethnobotany*. Timber Press, Portland OR, pp. 598–600.
117. Bailey, L.H. (1898) *Sketch of the evolution of our native fruits*. Macmillan, New York. Campbell, C. (2004) *Phylloxera: How wine was saved for the world*. HarperCollins, London.
118. Borneman, A.R., Schmidt, S.A., and Pretorius, I.S. (2013) At the cutting-edge of grape and wine biotechnology. *Trends in Genetics* 29: 263–71.
119. This, P., Lacombe, T., and Thomas, M.R. (2006) Historical origins and genetic diversity of wine grapes. *Trends in Genetics* 22: 511–19.
120. Walker, A.R., Lee, E., Bogs, J., McDavid, D.A., Thomas, M.R., Robinson, S.P. (2007) White grapes arose through the mutation of two similar and adjacent regulatory genes. *Plant Journal* 49: 772–85. Bowers, J.E., and Meredith, C.P. (1997) The parentage of a classic wine grape, Cabernet Sauvignon. *Nature Genetics* 16: 84–7. Bowers, J., Boursiquot, J.-M., This, P. Chu, K., Johansson, H., and Meredith, C. (1999) Historical genetics: The parentage of Chardonnay, Gamay, and other wine grapes of northeastern France. *Science* 285: 1562–5. Levadoux, L. (1956) Les populations sauvages et cultivées de *Vitis vinifera* L. *Annales de l'Amélioration des Plantes* 6: 59–117.
121. Myles, S. (2013) Improving fruit and wine: What does genomics have to offer? *Trends in Genetics* 29: 190–96.
122. Bronk Ramsey, C., Dee, M.W., Rowland, J.M., Higham, T.F.G., Harris, S.A., Brock, F., Quiles, A., Wild, E.M., Marcus, E.S., and Shortland, A.J. (2010) Radiocarbon-based chronology for Dynastic Egypt. *Science* 328: 1554–7.
123. Rackham, H. (1945) *Pliny: Natural History. Books 12–16*. Harvard University Press, Cambridge MA, Book 13, 23, 76.
124. Sider, D. (2005) *The Library of the Villa dei Papiri at Herculaneum*. J. Paul Getty Museum, Los Angeles.
125. *The Oxyrhynchus Papyri*. https://oxyrhynchus.web.ox.ac.uk (accessed 26 December 2023).
126. El-Abbadi, M. (1992) *Life and fate of the ancient Library of Alexandria*. UNESCO, Paris. Jochum, U. (1999) The Alexandrian Library and its aftermath. *Library History* 15: 5–12.
127. Heyerdahl, T. (1993) *The Ra Expeditions*. HarperCollins, London.
128. Okurut, T.O., Rijs. G.B.J., and van Bruggen, J.J.A. (1999) Design and performance of experimental constructed wetlands in Uganda, planted with *Cyperus papyrus* and *Phragmites mauritianus*. *Water Science and Technology* 40: 265–71.
129. Hunter, D. (1978) *Papermaking: The history and technique of an ancient craft*. Dover Publications, New York.
130. Rackham, H. (1945) *Pliny: Natural History. Books 12–16*. Harvard University Press, Cambridge MA, Book 13, 21, 70.
131. Fowler, B. (2001) *Iceman: Uncovering the life and times of a prehistoric man found in an alpine glacier*. University of Chicago Press, Chicago IL.
132. Oakley, K.P., Andrews, P., Keeley, L.H., and Clark, J.D. (1977) A reappraisal of the Clacton spearpoint. *Proceedings of the Prehistoric Society* 43: 13–30.
133. Farjon, A., and Filer, D. (2013) *An atlas of the world's conifers: An analysis of their distribution, biogeography, diversity and conservation status*. Brill, Leiden.
134. Günther, R.T. (1912) *Oxford gardens based upon Daubeny's popular guide to the*

Physick Garden of Oxford, with notes on the gardens of the Colleges and on the University Park. Parker & Son, Oxford, p. 189.

135. Hageneder, F. (2001) *The heritage of trees: History, culture and symbolism*. Floris Books, Edinburgh.
136. Hardy, R. (2012) *Longbow: A social and military history*. Haynes, Yeovil.
137. Hartzell, H.R. (1995) Yew and us: A brief history of the yew tree. In Suffness, M. (ed.) *Taxol: science and applications*. CRC Press, Boca Raton FL, pp. 27–34.
138. Edwards, H.J. (1917) *Caesar: De Bello Gallico. The Gallic War*. Heinemann, London, Book 6, 31.
139. Hageneder, F. (2007). *Yew: A history*. Sutton Publishing, Stroud. Frohne, D., and Pfander, H.J. (2005) *Poisonous plants: A handbook for doctors, pharmacists, toxicologists, biologists and veterinarians*. Manson Publishing, London.
140. Goodman, J., and Walsh, V. (2006) *The story of taxol: Nature and politics in the pursuit of an anti-cancer drug*. Cambridge University Press, Cambridge.
141. Linnaeus, C. (1753) *Species plantarum*. Laurentii Salvii, Holmiae, p. 492.
142. Lindley, J. (1820). *Rosarum Monographia, or A botanical history of roses*. Printed for J. Ridgeway, London. Gandoger, M. (1892–93). *Monographia Rosarum Europae et Orientis*. Ballière, Paris.
143. Crépin, F. (1893) L'obsession de l'individu dans l'étude des Roses. *Bulletin de la Société royale de Botanique de Belgique* 32: 52–5. Crépin, F. (1894) Rosae hybrida: Études sur les Roses hybrides. *Bulletin de la Société royale de Botanique de Belgique* 33: 7–149.
144. Erlanson, E.W. (1929) Chromosome studies and evolution in the genus *Rosa*. *Bulletin de la Jardin Botanique de Natura Belgique* 37: 45–52. Gustafsson, A., and Håkansson, A. (1942) Meiosis in some *Rosa*-hybrids. *Botaniska Notiser* 1942: 331–43. Rowley, G.D. (1967) Chromosome studies and evolution in the genus *Rosa*. *Bulletin de la Jardin Botanique de Natura Belgique* 37: 4552.
145. Rehder, A. (1940) *Manual of cultivated trees and shrubs hardy in North America*, 2nd edn. Macmillan, New York. Matthews, V.A. (1995) *Rosa* Linnaeus. In Cullen, J., Alexander, J.C.M., Brady, A., Bickell, C.D., Green,P.S., Heywood, V.H., Jörgensen, P.M., Jury, S.L., Knees, S.G., Leslie, A.C., Matthews, V.A., Robson,N.K.B., Walters, S.M., Wijnands, D.O., and Yeo, P.F. (eds) *The European garden Flora: A manual for the identification of plants cultivated in Europe, both out-of-doors and under glass*, Cambridge University Press, Cambridge, pp. 358–79.
146. Stearn, W.T. (1965) The five brethren of the rose: An old botanical riddle. *Huntia* 2: 180–84.
147. *Rosa chinensis*, *R. foetida*, *R. gallica*, *R. gigantea*, *R. moschata*, *R. multiflora* and *R. wichurana*. Wylie, A.P. (1954). The history of the garden rose. *Journal of the Royal Horticultural Society* 79: 555–71.
148. Bruneau, A., Starr, J.R., and Joly, S. (2007) Phylogenetic relationships in the genus *Rosa*: New evidence from chloroplast DNA sequences and an appraisal of current knowledge. *Systematic Botany* 32: 366–78. Matsumoto, S., Kouchi, M., Yabuki, J., Kusunoki, M., Ueda, Y., and Fukui, H. (1998) Phylogenetic analyses of the genus *Rosa* using the *mat*K sequence: Molecular evidence for the narrow genetic background of modern roses. *Scientia Horticulturae* 77: 73–82.
149. Young, N. (1971) *The complete rosarian: The development, cultivation and reproduction of rose*. Hodder & Stoughton, London.

150. Cited in the entry for carnations by Thornton, R.J. (1807) *New illustration of the sexual system of Carolus von Linnaeus: and the Temple of Flora, or garden of nature.* Published for Author, London.
151. Weir, A. (1998). *Lancaster and York: The Wars of the Roses.* Vintage, London.
152. Hort, A. (1926) *Theophrastus: Enquiry into plants*, Vol. 2. William Heinemann, London, ch. 6, 8.
153. Kuniholm, P.I. (2001) Dendrochronology and other applications of tree-ring studies in archaeology. In Brothwell, D.R., and Pollard, A.M. (eds) *The handbook of archaeological science.* John Wiley, London, pp. 1–11.
154. Fritz, H.C. (1991) *Reconstructing large-scale climatic patterns from tree-ring data.* University of Arizona Press, Tucson AZ. Stokes, M.A., and Smiley, T.L. (1996) *An introduction to tree-ring dating.* University of Arizona Press, Tucson AZ.
155. Friedrich, M., Remmele, S., Kromer, B., Hofmann, J., Spurk, M, Kaiser, K.F., Orcel, C., and Küppers, M. (2004) The 12,460-year Hohenheim oak and pine tree-ring chronology from Central Europe – a unique annual record for radiocarbon calibration and paleoenvironment reconstructions. *Radiocarbon* 46: 1111–22.
156. Godwin, H. (1956) *The history of the British flora.* Cambridge University Press, Cambridge.
157. Farjon, A., and Filer, D. (2013) *An atlas of the world's conifers: An analysis of their distribution, biogeography, diversity and conservation status.* Brill, Leiden.
158. Penney, D., and Green, D.I. (c.2011) *Fossils in amber: Remarkable snapshots of prehistoric forest life.* Siri Scientific Press, Manchester. Poinar, G., and Poinar, R. (1999) *The amber forest: A reconstruction of a vanished world.* Princeton University Press, Princeton NJ.
159. Williams, J. (1935) English mercantilism and Carolina naval stores, 1705–1776. *The Journal of Southern History* 1: 169–85. Malone, J.J. (1964) *Pine trees and politics: The naval stores and forest policy in colonial New England, 1691–1775.* Longmans, London.
160. USDA (1935) *A naval stores handbook dealing with the production of pine gum or oleoresin.* United States Department of Agriculture, Washington DC.
161. Williams, M. (2006) *Deforesting the earth: From prehistory to global crisis. An abridgement.* University of Chicago Press, Chicago IL.
162. Maxwell, G. (1959) *A reed shaken by the wind: A journey through the unexplored marshlands of Iraq.* Longmans, London. Thesiger, W. (1964) *The marsh Arabs.* Longmans, London. Young, G. (2009) *Return to the marshes: Life with the marsh Arabs of Iraq.* Faber & Faber, London.
163. Cope, T., and Gray, A. (2009) *Grasses of the British Isles.* Botanical Society of the British Isles, London.
164. Ridley, H.N. (1930) *The dispersal of plants throughout the world.* Reeve, Ashford.
165. Mazoyer, M., and Roudart, L. (2006) *A history of world agriculture from the neolithic age to the current crisis.* Earthscan, London.
166. Curtis J., Richardson, C.J., Reiss, P., Hussain,, N.A., Alwash, A.J., Pool, D.J. (2005) The restoration potential of the Mesopotamian Marshes of Iraq. *Science* 307: 1307–11.
167. Zuazo, V.H.D., and Pleguezuelo, C.R.R. (2009) Soil-erosion and runoff prevention by plant covers: a review. In Lichtfouse, E., Navarrete, M., Debaeke, P., Souchère, V., and Alberola, C.(eds) *Sustainable agriculture.* Springer, Dordrecht,

pp. 785–811. Pimentel, D., Harvey, C., Resosudarmo, P., Sinclair, K., Kurz, D., McNair, M., Crist, S., Shpritz, L., Fitton, L., Saffouri, R., and Blair, R. (1995) Environmental and economic costs of soil erosion and conservation benefits. *Science* 267: 1117–23.

168. Vymazal, J. (2010) Constructed wetlands for wastewater treatment. *Water* 2: 530–49.
169. Lambert, J.M., Jennings, J.N., Smith, C.T., Green, C., and Hutchinson J.N. (1960) *The making of the Broads: A reconsideration of their origin in the light of new evidence*. Royal Geographical Society, London.
170. Mitsch, W.J. (2014) When will ecologists learn engineering and engineers learn ecology? *Ecological Engineering* 65: 9–14. Nama, A.H., Alwan, I.A. and Pham, Q.B. (2024) Climate change and future challenges to the sustainable management of the Iraqi marshlands. *Environmental Monitoring and Assessment* 196: 35.
171. International Energy Agency (2011) *Technology roadmap: Biofuels for transport*. International Energy Agency, Paris. Jeswani, K.K., Chilvers, A., and Azapagic, A. (2020) Environmental sustainability of biofuels: A review. *Proceedings of the Royal Society A: Mathematical, Physical and Engineering Sciences* 476: 20200351.
172. Mariani, C., Cabrini, R., Danin, A., Piffanelli, P., Fricano, A., Gomarasca, S., Dicandilo, M., Grassi, F., and Soave, C. (2010) Origin, diffusion and reproduction of the giant reed (*Arundo donax* L.): A promising weedy energy crop. *Annals of Applied Biology* 157: 191–202. Naik, S.N., Goud, V.V., Rout, P.K., and Dalai, A.K (2010) Production of first and second generation biofuels: A comprehensive review. *Renewable and Sustainable Energy Reviews* 14: 578–97.
173. Kreiner, J. (2021) *Legions of pigs in the Early Medieval West*. Yale University Press, New Haven CT.
174. Hewitt, G. (2000) The genetic legacy of Quaternary ice ages. *Nature* 405: 907–13.
175. Petit, R.J., Csaikl, U.M., Bordács, S., Burg, K., Coart, E., Cottrell, J., van Dam, B., Deans, J.D., Lapegue-Dumolin, S., Fineschi, S., Finkeldey, R., Gillies, A.C.M., Glaz, I., Goicoechea, P.G., Jensen, J.S., Konig, A.O., Lowe, A.J., Madsen, S.F., Matyas, G., Munro, R.C., Olalde, M., Pemonge, M.H., Popescu, F., Slade D., Tabbener, H., Taurchini, D., and van Dam, B. (2002) Chloroplast DNA variation in European white oaks: Phylogeography and patterns of diversity based on data from over 2600 populations. *Forest Ecology and Management* 156: 5–26.
176. Hemery, G., and Simblet, S. (2014) *The New Sylva: A discourse of forest & orchard trees for the twenty-first century*. Bloomsbury, London, pp. 202–15.
177. Marsden, P. (2003) *Sealed by time: The loss and recovery of the Mary Rose*. The Archaeology of the *Mary Rose*, Vol. 1. Mary Rose Trust, Portsmouth.
178. Rackham, O. (1995). *Trees and woodlands in the British landscape: The complete history of Britain's trees, woods and hedgerows*. Weidenfeld & Nicolson, London.
179. Mabey, R. (2006). *Fencing paradise: The uses and abuses of plants*. Transworld Publishers, London.
180. Diringer, D. (1982) *The book before printing: Ancient, medieval and oriental*. Dover Publications, New York.
181. Rackham, O. (1995). *Trees and woodlands in the British landscape: The complete history of Britain's trees, woods and hedgerows*. Weidenfeld & Nicolson, London.
182. Mason, S. (1995) Acornutopia? Determining the role of acorns in past subsistence. In Wilkins, J., Harvey, D., and Dobson, M. (eds) *Food in antiquity*. University of

Exeter Press, Exeter, pp. 12–24. Pearman, G. (2005). Nuts, seeds and pulses. In Prance, G.T., and Nesbitt, M. (eds) *The cultural history of plants*. Routledge: New York, pp. 133–52.
183. Dzhangaliev, A.D. (2010) The wild apple tree of Kazakhstan. In Janick, J. (ed.) *Horticultural Reviews: Wild apple and fruit trees of Central Asia*, Vol. 29. John Wiley, Oxford.
184. Cornille, A., Giraud, T., Smulders, M.J.M., Roldán-Ruiz, I., and Gladieux, P. (2014) The domestication and evolutionary ecology of apples. *Trends in Genetics* 30: 57–65.
185. Cornille, A., Giraud, T., Smulders, M.J.M., Roldán-Ruiz, I., and Gladieux, P. (2014) The domestication and evolutionary ecology of apples. *Trends in Genetics* 30: 57–65.
186. McFadden, C. (2008) *Pepper: The spice that changed the world*. Absolute Press, Bath, p. 78.
187. Hort, A. (1926) *Theophrastus: Enquiry into plants*, Vol. 2. William Heinemann, London, ch. 9, 20.
188. Miller, I.J. (1969) *The spice trade of the Roman Empire*. Oxford University Press, Oxford. Thapar, R. (1992) Black gold: South Asia and the Roman maritime trade. *South Asia: Journal of South Asian Studies* 15: 1–27.
189. Prange, S.R. (2011) 'Measuring by the bushel': Reweighing the Indian Ocean pepper trade. *Historical Research* 84: 212–35.
190. Purseglove, J.W., Brown, E.G., Green, C.L., and Robbins, S.R.J. (1981) *Spices*, Vol. 1. Longman Scientific and Technical, London.
191. Freedman, P. (2005) Spices and late-Medieval European ideas of scarcity and value. *Speculum* 80: 1209–27.
192. Johns, C. (2010) *The Hoxne late Roman treasure: Gold, jewellery and silver plate*. British Museum Press, London.
193. Cook, H.J. (2007) *Matters of exchange: Commerce, medicine, and science in the Dutch Golden Age*. Yale University Press, New Haven CT.
194. Accum, F. (1820) *A treatise on adulteration of food*. Longman, Hurst, Rees, Orme & Brown, London, p. 285.
195. Accum, F. (1820) *A treatise on adulteration of food*. Longman, Hurst, Rees, Orme & Brown, London, p. 286.
196. Norwich, J.J. (1990) *Byzantium: The early centuries*. Penguin, London, p. 134.
197. *Genesis* 2:20.
198. Gulick, C.B. (1930) *Athenaeus: The Deipnosophists. Book IX*. Harvard University Press, Cambridge MA, p. 183.
199. Stolarczyk, J., and Janick, J. (2011) Carrot: History and iconography. *Chronica Horticulturae* 51: 13–18.
200. Culpeper, N. (1653) *The English physitian enlarged*. Printed by Peter Cole, London, p. 56.
201. Just, B., Santos, C. F., Yandell, B., and Simon, P. (2009) Major QTL for carrot color are positionally associated with carotenoid biosynthetic genes and interact epistatically in a domesticated × wild carrot cross. *Theoretical and Applied Genetics* 119: 1155–69.
202. Surles, R.L., Weng, N., Simon, P.W., Tanumihardjo, S.A. (2004) Carotenoid

profiles and consumer sensory evaluation of speciality carrots. *Journal of Agricultural and Food Chemistry* 52: 3417–21.
203. Brothwell, D.R., and Brothwell, P. (1969) *Food in antiquity: A survey of the diet of early peoples*. Thames & Hudson, London.
204. Iorizzo, M., Genalik, D.A., Ellison, S.L., Grzebelus, D., Cavagnaro, P.F., Allender, C., Brunet, J., Spooner, D.M., van Deynze A., and Simon, P.W. (2013) Genetic structure and domestication of carrot (*Daucus carota* subsp. *sativus*) (Apiaceae). *American Journal of Botany* 100: 930–38.
205. Vavilov, N.I. (1992) *Origin and geography of cultivated plants*. Cambridge University Press, New York, pp. 337–40.
206. Mackevic, V. (1932) *The carrot of Turkey*. WACCHNIL, Institute of Plant Breeding, Leningrad, Russia. Zagorodskikh, P. (1939) New data on the origin and taxonomy of the cultivated carrot. *Comptes Rendus (Doklady), Academy of Science, USSR* 25: 522–5. Banga, O. (1963) Origin and domestication of the western cultivated carrot. *Genetica Agraria* 17: 357–70.
207. Banga, O. (1957) Origin of the European cultivated carrot: The development of the original European carrot material. *Euphytica 6*: 64–76. Banga, O. (1963) Origin and domestication of the western cultivated carrot. *Genetica Agraria 17*: 357–70.
208. Zeven, A.C., and Brandenburg, W.A. (1986) Use of paintings from the sixteenth to nineteenth centuries to study the history of domesticated plants. *Economic Botany 40*: 397–408. Stolarczyk, J., and Janick, J. (2011) Carrot: History and iconography. *Chronica Horticulturae* 51: 13–18.
209. Vilmorin, M. (1859) L'hérédité dans les végétaux. In Vilmorin, M., *Notice sur l'amélioration des plantes par la semis*. Librairie Agricole, Paris, pp. 5–29. Small, E. (1978) Numerical taxonomic analysis of *Daucus carota* complex. *Canadian Journal of Botany* 56: 248–76. Heywood, V.H. (1983) Relationships and evolution in the *Daucus carota* complex. *Israel Journal of Ecology and Evolution* 32: 51–65.
210. Banga, O. (1957) The development of the original European carrot material. *Euphytica* 6: 64–76. Iorizzo, M., Genalik, D.A., Ellison, S.L., Grzebelus, D., Cavagnaro, P.F., Allender, C., Brunet, J., Spooner, D.M., van Deynze A., and Simon, P.W. (2013) Genetic structure and domestication of carrot (*Daucus carota* subsp. *sativus*) (Apiaceae). *American Journal of Botany* 100: 930–38.
211. Darwin, T. (1996) *The Scots Herbal: The plant lore of Scotland*. Mercat Press, Edinburgh.
212. Ginn, F. (2012) Dig for Victory! New histories of wartime gardening in Britain. *Journal of Historical Geography* 38: 294–305.
213. Edwards, H.J. (1917) *Caesar: De Bello Gallico. The Gallic War*. Heinemann, London, Book 5, 14. Foster, S.M. (1996) *Picts, Gaels and Scots*. Batsford, London.
214. Pyatt, F.B., Beaumont, E.H., Lacy. D., Magilton, I.R., and Buckland, P.C. (1991) Non isatis sed vitrum or, the colour of Lindow Man. *Oxford Journal of Archaeology* 10: 61–73.
215. Van der Veen, M., Hall, A.R., and May, J. (1993) Woad and the Britons painted blue. *Oxford Journal of Archaeology* 12: 367–71. Hall, A.R. (1992) Archaeological records of woad (*Isatis tinctoria* L.) from medieval England and Ireland. *Beitruge zur Waidtagung* 4: 21–2. Tomlinson, P. (1985) Use of vegetative remains in the identification of dyeplants from waterlogged ninth–tenth century deposits at York. *Journal of Archaeological Science* 12: 269–83.

216. Hurry, J.B. (1930) *The woad plant and its dye*. Oxford University Press, Oxford. Brown, M.P. (2003) *The Lindisfarne Gospels. Society, spirituality and the scribe*. University of Toronto Press, Toronto.
217. Elizabeth I (1585) *By the Queene. A proclamation against the sowing of woade*. Christopher Barker, London.
218. Gerard, J. (1597) *The herball or generall historie of plantes*. John Norton, London, p. 395.
219. Davis, M. (2001) *Late Victorian holocausts: El Niño famine and the making of the Third World*. Verso, London.
220. Williams, G. (1999) *The prize of all the oceans*. Viking, New York.
221. Lind, J. (1753) *A treatise of the scurvy*. Sands, Murray, and Cochran, Edinburgh, pp. 180–239.
222. Lind, J. (1753) *A treatise of the scurvy*. Sands, Murray, and Cochran, Edinburgh, pp. 191–3.
223. Hess, A.S. (1920) *Scurvy past and present*. J.B. Lippencoer, Philadelphia PA.
224. Calabrese, F. (2002) Origin and history. In Dugo, G. and di Giacomo, A. *Citrus: The genus* Citrus. Taylor & Francis, London, pp. 1–15. Garcia-Lor, A., Curk, F., Snoussi-Trifa, H., Morillon, R., Ancillo, G., Luro, F., Navarro, L., and Ollitrault P. (2013) A nuclear phylogeny: SNPs, indels and SSRs deliver new insights into the relationships in the 'true citrus fruit trees' group (Citrinae, Rutaceae) and the origin of cultivated species. *Annals of Botany* 111: 1–19. Wu, G.A., Prochnik, S., Jenkins, J., Salse, J., Hellsten, U., Murat, F., Perrier, X., Ruiz, M., Scalabrin, S., Terol, J., Takita, M.A., Labadie, K., Poulain, J., Couloux, A., Jabbari, K., Cattonaro, F., del Fabbro, C., Pinosio, S., Zuccolo, A., Chapman, J., Grimwood, J., Tadeo, F.R., Estornell, L.H., Muñoz-Sanz, J.V., Ibanez, V., Herrero-Ortega, A., Aleza, P., Pérez-Pérez, J., Ramón, D., Brunel, D., Luro, F., Chen, C., Farmerie, W.G., Desany, B., Kodira, C., Mohiuddin, M., Harkins, T., Fredrikson, K., Burns, P., Lomsadze, A., Borodovsky, M., Reforgiato, G., Freitas-Astúa, J., Quetier, F., Navarro, L., Roose, M., Wincker, P., Schmutz, J., Morgante, M., Machado, M.A., Talon, M., Jaillon, O., Ollitrault, P., Gmitter, F., and Rokhsar, D. (2014) Sequencing of diverse mandarin, pummelo and orange genomes reveals complex history of admixture during citrus domestication. *Nature Biotechnology* 32: 656–62.
225. Calabrese, F. (2002) Origin and history. In Dugo, G., and di Giacomo, A., *Citrus: The genus* Citrus. Taylor & Francis, London, pp. 1–15. Gmitter, F.G. and Hu, X. (1990) The possible role of Yunnan, China, in the origin of contemporary citrus species (Rutaceae). *Economic Botany* 44: 267–77. Ramón-Laca, L. (2003) The introduction of cultivated citrus to Europe via northern Africa and the Iberian Peninsula. *Economic Botany* 57: 502–14.
226. Díaz, B. (1963) *The conquest of New Spain*. Penguin Books, London.
227. Rea, J. (1676) *Flora, Ceres, Pomona*. Printed by T.N. for George Marriott, London.
228. Hamilton, W. (1796) A short account of several gardens near London, with remarks on some particulars wherein they excel, or are deficient, upon a view of them in December 1691. Communicated to the Society by the Reverend Dr. Hamilton, Vice President, from an original manuscript in his possession. *Archaeologia* 12: 181–92.
229. Sackman, D. (2005) *Orange empire: California and the fruits of Eden*. University of California Press, Berkeley.

230. Cogan, T. (1636) *The haven of health. Chiefly gathered for the comfort of students, and consequently of all those that have a care of their health.* Printed by Anne Griffin, London, p. 125.
231. Donkin, R.A. (2003) *Between east and west: The Moluccas and the traffic in spices up to the arrival of Europeans.* American Philosophical Society, Philadelphia PA.
232. Ellacombe, H.N. (1878) The plant-lore and garden-craft of Shakespeare. William Pollard, London.
233. Gerard's *Herball* (1597: 1353–4) provides detailed woodcuts of the nutmeg tree but he clearly had not seen the plant; the leaves have serrated rather than entire margins. A notable exception was Garcia da Orta, who was familiar with the tree as an introduced species in Goa, where he was based for nearly three decades. Markham, C. (1913) *Colloquies of the simples and drugs of India by Garcia da Orta.* Henry Sotheran, London, p. 273.
234. Donkin, R.A. (2003) *Between east and west: The Moluccas and the traffic in spices up to the arrival of Europeans.* American Philosophical Society, Philadelphia PA.
235. Markham, C. (1913) *Colloquies of the simples and drugs of India by Garcia da Orta.* Henry Sotheran, London, p. 273.
236. Boxer, C.R. (1965) *The Dutch seaborne empire: 1600–1800.* Hutchinson, London.
237. Donkin, R.A. (2003) *Between east and west: The Moluccas and the traffic in spices up to the arrival of Europeans.* American Philosophical Society, Philadelphia PA.
238. The VOC eventually went bankrupt in the early 1800s and its nutmeg monopoly was taken over by the Dutch government. Cook, H.J. (2007) *Matters of exchange. Commerce, medicine, and science in the Dutch Golden Age.* Yale University Press, New Haven CT.
239. Wallace, A.R. (1869) *The Malay Archipelago: The land of the Orang-Utan, and the Bird of Paradise.* Macmillan, London, p. 295.
240. Ovington, J. (1696) *A voyage to Suratt in the year 1689.* Printed for Jacob Tonson, London, pp. 225–6.
241. Wallace, A.R. (1869) *The Malay Archipelago. The land of the Orang-Utan, and the Bird of Paradise.* Macmillan, London, pp. 295–7.
242. Grove, R.H. (1995) *Green imperialism: Colonial expansion, tropical island Edens and the origins of environmentalism, 1660–1860.* Cambridge University Press, Cambridge, pp. 168–263. Spary, E.C. (2005) Of nutmegs and botanists: The colonial cultivation of botanical identity. In Schiebinger, L., and Swan, C. *Colonial botany. Science., commerce, and politics in the early modern world.* University of Pennsylvania Press, Philadelphia PA, pp. 187–203.
243. Bovill, E.W. (1968) The Madre de Dios. *The Mariner's Mirror* 54: 129–52.
244. Liu, X., and Shaffer, L.N. (2007) *Connections across Eurasia: Transportation, communication, and cultural exchange on the Silk Roads.* McGraw–Hill, Boston MA.
245. Feltwell, J. (1990) *The story of silk.* St Martin's Press, London.
246. Arunkumar, K.P., Metta, M., and Nagaraju, J.H. (2006) Molecular phylogeny of silkmoths reveals the origin of domesticated silkmoth, *Bombyx mori* from Chinese *Bombyx mandarina* and paternal inheritance of *Antheraea proylei* mitochondrial DNA. *Molecular Phylogenetics and Evolution* 40: 419–27.
247. Dewhurst, H.W. (1836) A familiar treatise on the natural history and management of the Phalaena *Bombyx mori* or common silk-worm. J. Saunders, London.

248. Milne, R.I., and Abbott, R.J. (2002) The origin and evolution of tertiary relict floras. *Advances in Botanical Research* 38: 281–314.
249. Wang, Z. (1987). *Sericulture in Ancient China's technology and science*. Institute of the History of Natural Sciences Chinese Academy of Sciences, Beijing.
250. Thorley, J. (1971) The silk trade between China and the Roman Empire at its height, circa A.D. 90–130. *Greece and Rome* 18: 71–80.
251. Foster, G.R. (1831) *A treatise on the origin, progressive improvement, and present state of the silk manufacture*. Longman, Rees, Orme, Brown, and Green, London.
252. Mola, L. (2000) *The silk industry of Renaissance Venice*. Johns Hopkins University Press, Baltimore MD.
253. MacGregor, N. (2012) *A history of the world in 100 objects*. Penguin Books, London, pp. 271–5.
254. Roach, F.A. (1985) *Cultivated fruits of Britain: Their origins and history*. Blackwells, Oxford.
255. Cobb, J.H. (1831) *A manual containing information respecting the growth of the mulberry tree, with suitable directions for culture of silk*. Carter, Hendee & Co. Ryland, E.H. (1939) America's 'Multicaulis Mania'. *William and Mary Quarterly Historical Magazine* 19: 25–33. Cole, A.H. (1926) Agricultural crazes: A neglected chapter in American economic history. *American Economic Review* 16: 627–32.
256. Anonymous (1832) On the management of the silk-worm; and on growing silk, as means of bettering the condition of the labouring classes. Part 3. The *Horticultural Register and General Magazine* 1832: 412–19. Rhind, W. (1877) *A history of the vegetable kingdom*. Blackie & Son. Anonymous (1854) *The art of breeding, rearing, and keeping silkworms, so that they may prove profitable to those that undertake their management*. W. Mason, London.
257. Parkinson, J. (1629) *Paradisi in sole*. Printed by Humfrey Lownes & Robert Young, London, p. 596.
258. Sánchez, M.D. (2002) *Mulberry for animal production. FAO Animal Production and Health Paper 147*. FAO, Rome. Anonymous (1825) *The art of rearing silk-worms. Translated from the work of Count Dandolo*. John Murray, London.
259. Wisconsin Historical Society (2003) A shorte and briefe narration (Cartier's Second Voyage), 1535–1536. Document No. AJ-027. Wisconsin Historical Society Digital Library and Archives, p. 68.
260. Townsend, T. (1738) *The history of the conquest of Mexico by the Spaniards. Translated into English from the original Spanish of Don Antonio de Solis*, Vol. 1. Printed for John Osborn, London, p. 417.
261. Coles, E. (1717) *An English dictionary, explaining the difficult terms that are used in divinity, husbandry, physick, philosphy, law, navigation, mathematicks, and other arts and sciences*. Printed by S. Collins, London.
262. Chase, M.W., Knapp, S., Cox, A.V., Clarkson, J.J., Butsko, Y., Joseph, J., Savolainen, V., and Parokonny, A.S. (2003) Molecular systematic, GISH and the origin of the hybrid taxa in *Nicotiana* (Solanaceae). *Annals of Botany* 92: 107–27.
263. Hancocke, J. (1672) *King James his counterblast to tobacco. To which is added a learned discourse proving that tobacco is a procuring cause of the scurvy*. Printed for John Hancock, London, p. 12.
264. Hariot, T. (1590) *A briefe and true report of the new found land of Virginia*. Typis Ioannis Wecheli, sumptibus vero Theodori de Bry, Francoforti ad Moenum: p. 16.

Arianrhod, R. (2019) *Thomas Harriot: A life in science*. Oxford University Press, New York.
265. The Tobacco Atlas (2023) *The Tobacco Atlas*. https://tobaccoatlas.org (accessed 28 December 2023). World Health Organization (2009) *Global health risks: Mortality and burden of disease attributable to selected major risks*. World Health Organization, Geneva. World Health Organization (2023) Tobacco. www.who.int/health-topics/tobacco#tab=tab_1 (accessed 28 December 2023).
266. Nutt, D., King, L.A., Saulsbury, W., and Blakemore, C. (2007) Development of a rational scale to assess the harm of drugs of potential misuse. *The Lancet* 369: 1047–53.
267. World Health Organization (2009) *Global health risks: Mortality and burden of disease attributable to selected major risks*. World Health Organization, Geneva. The Tobacco Atlas (2023) *The Tobacco Atlas*. https://tobaccoatlas.org (accessed 28 December 2023).
268. The Tobacco Atlas (2023) *The Tobacco Atlas*. https://tobaccoatlas.org (accessed 28 December 2023). Food and Agriculture Organization of the United Nations (2023) FAOSTAT http://faostat.fao.org (accessed 28 December 2023).
269. Zirkle, C. (1935) *The beginnings of plant hybridization*. University of Pennsylvania, Philadelphia PA.
270. Fraley, R.T., Rogers, S.G., Horsch, R.B., Sanders, P.R., Flick, J.S., Adams, S.P., Bittner, M.L., Brand, L.A., Fink, C.L., Fry, J.S., Galluppi, G.R., Goldberg, S.B., Hoffmann, N.L. and Woo, S.C. (1983) Expression of bacterial genes in plant cells. *Proceedings of the National Academy of Sciences USA* 80: 4803–7.
271. Mackay, C. (1841) *Memoirs of extraordinary popular delusions and the madness of crowds*. Richard Bentley, London.
272. Dubos, R.J. (1960) Tulipomania and the benevolent virus. *Garden Journal of the New York Botanical Garden* 10: 41–3, 66–7.
273. Garber, P.M. (1989) Tulipmania. *Journal of Political Economy* 97: 535–60. Goldgar, A. (2007) *Tulipmania: Money, honor, and knowledge in the Dutch Golden Age*. University of Chicago Press, Chicago IL. Thompson, E. (2007) The tulipmania: Fact or artifact? *Public Choice* 130: 99–114.
274. Hall, D. (1940) *The genus Tulipa*. Royal Horticultural Society, London.
275. Salzmann, A. (2000) The Age of Tulips confluence and conflict in Early Modern consumer culture (1550–1730). In Quataert, D., *Consumption studies and the history of the Ottoman Empire, 1550–1922*. State University of New York Press, Albany NY, pp. 83–106.
276. Delaunay, P. (1923–24) Pierre Beon Naturaliste. *Bulletin de la Société d'Agriculture, Sciences et Arts de la Sarthe* 41: 13–39. Deschamps, L. (1887) Pierre Belon: naturaliste et explorateur. *Revue de Géographie* 21: 433–40.
277. Forster, E.S. (1927) *The Turkish letters of Ogier Ghiselin de Busbecq, Imperial Ambassador at Constantiople, 1554–1562*. Clarendon Press, Oxford.
278. Gesner, C. (1561) *De hortis Germaniae cum Descriptione Tulipae Turcarum*. Excudebat Iosias Rihelius, Argentorati.
279. Egmond, F. (2010) *The world of Carolus Clusius: Natural history in the making, 1550–1610*. Pickering & Chatto, London. Ogilvie, B.W. (2006) *The science of describing: Natural history in Renaissance Europe*. University of Chicago Press, Chicago IL.

280. Collins, M.D., Wasmund, L.M., and Bosland, P.W. (1995) Improved method for quantifying capsaicinoids in *Capsicum* using high-performance liquid chromatography. *HortScience* 30: 137–9.
281. Bosland, P.W., Coon, D., and Reeves, G. (2012) 'Trinidad Moruga Scorpion' pepper is the world's hottest measured chile pepper at more than 2 million Scoville heat units. *HortTechnology* 22: 534–8.
282. Rozin, P., and Schiller, D. (1980). The nature and acquisition of a preference for chili pepper by humans. *Motivation and Emotion* 4: 77–101.
283. Knapp, S. (2007) Some like it hot. *Science* 315: 946–7.
284. Perry, L., Dickau, R., Zarrillo, S., Holst, I., Pearsall, D.M., Piperno, D.R., Berman, M.J., Cooke, R.G., Rademaker, K., Ranere, A.J., Raymond, J.S., Sandweiss, D.H., Scaramelli, F., Tarble, K., and Zeidler, J.A. (2007) Starch fossils and the domestication and dispersal of chili peppers (*Capsicum* spp. L.) in the Americas. *Science* 315: 986–8. Perry, L., and Flannery, K.V. (2007) Pre-Columbian use of chili peppers in the Valley of Oaxaca, Mexico. *Proceedings of the National Academy of Sciences USA* 104: 11905–9.
285. Markham, C.R. (1893) *The journal of Christopher Columbus (during his first voyage, 1492–1493), and documents relating to the voyages of John Cabot and Gaspar Corte Real.* Hakluyt Society, London, p. 164.
286. Major, R.H. (1870) *Select letters of Christopher Columbus, with other original documents, relating to his four voyages to the New World.* Hakluyt Society, London, p. 68.
287. Gerard, J. (1597) *The herball or generall historie of plantes.* John Norton, London, pp. 292–3.
288. Whitehead, N.L., and Harbsmeier, M. (2008) *Hans Staden's true history: An account of cannibal captivity in Brazil.* Duke University Press, Durham NC.
289. Tootal, A. (1874) *The captivity of Hans Stade of Hesse, in A.D. 1547–1555, among the wild tribes of eastern Brazil.* Hakluyt Society, London, pp. 154, 166.
290. Tewksbury, J.J., and Nabhan, G.P. (2001) Directed deterrence by capsaicin in chillies. *Nature* 412: 403–4. Tewksbury, J.J., Levey, D.J., Huizinga, M., Haak, D.C., and Traverse, A. (2008) Costs and benefits of capsaicin-mediated control of gut retention in dispersers of wild chilies. *Ecology* 89: 107–17. Levey, D.J., Tewksbury, J.J., Cipollini, M.L., and Carlo, T.A. (2006) A field test of the directed deterrence hypothesis in two species of wild chilli. *Oecologia* 150: 61–8.
291. Jensen, P.G., Curtis, P.D., Dunn, J.A., Austic, R.E., and Richmond, M.E. (2003) Field evaluation of capsaicin as a rodent aversion agent for poultry feed. *Pest Management Science* 59: 1007–15.
292. Haggis, A.W. (1942) Fundamental errors in the early history of *Cinchona*. *Bulletin of History of Medicine* 10: 586–92. Jaramillo-Arango, J. (1949) A critical review of the basic facts in the history of *Cinchona*. *Journal of the Linnean Society of London* 53: 272–309.
293. Andersson, L. (1998) *A revision of the genus Cinchona (Rubiaceae–Cinchoneae).* New York Botanical Garden, New York.
294. Markham, C. (1880) *Peruvian bark: A popular account of the introduction of Chinchona cultivation into British India 1860–1880.* J. Murray, London.
295. Anonymous (1931) 'Introduction of Cinchona to India', *Kew Bulletin of Miscellaneous Information* 1931: 113–17.

296. Hooker, W.D. 1839. *Inaugural dissertation upon the Cinchonas, their history, uses and effects*. Glasgow University Press, Glasgow.
297. Spruce, R. (1908) *Notes of a botanist on the Amazon and Andes: Being records of travel on the Amazon and its tributaries as also to the Cataracts of the Orinoco during the years 1849–1864, 1817–1893*, Vol. II. Macmillan, London.
298. Letter from Robert Cross to Secretary of State for India, dated 9 November 1861. PBB. (1852–1863) Parliamentary Blue Book, Cinchona. Stationery Office, London.
299. Purseglove, J.W. (1968) *Tropical crops: Dicotyledons 2*. Longmans, London.
300. Henderson, J.S., Joyce, R.A., Hall, G.R., Hurst, W.J., and McGovern, P.E. (2007) Chemical and archaeological evidence for the earliest cacao beverages. *Proceedings of the National Academy of Sciences USA* 104: 18937–40.
301. Crown, P.L., and Hurst, W.J. (2009) Evidence of cacao use in the pre-Hispanic American Southwest. *Proceedings of the National Academy of Sciences USA* 106: 2110–13.
302. McNeil, C. (2006) *Chocolate in Mesoamerica: A cultural history of cacao*. University of Florida Press, Gainesville FL.
303. Motamayor, J.C., Lachenaud, P., da Silva e Mota, J.W., Loor, R., Kuhn, D.N., Brown, J.S., and Schnell, R.J. (2008) Geographic and genetic population differentiation of the Amazonian chocolate tree (*Theobroma cacao* L.). *PLoS ONE* 3: e3311. Clement, C.R., de Cristo-Araújo, M., d'Eeckenbrugge, G.C., Alves Pereira, A., and Picanço-Rodrigues, D. (2010) Origin and domestication of native Amazonian crops. *Diversity* 2: 72–106. Thomas, E., van Zonneveld, M., Loo, J., Hodgkin, T., Galluzzi, G., van Etten, J., and Fuller, D.Q. (2012). Present spatial diversity patterns of *Theobroma cacao* L. in the Neotropics reflect genetic differentiation in Pleistocene refugia followed by human-influenced dispersal. *PLoS ONE* 7: e47676.
304. From Girolamo Benzoni's *La Historia del Mundo Novo* (1575), cited in Coe, S.D., and Coe, M.D. (1996) *The true history of chocolate*. Thames & Hudson, London.
305. McEwan, G.F. (2008) *The Incas: New perspectives*. W.W. Norton, London.
306. Spooner, D.M., McLean, K., Ramsay, G., Waugh, R., and Bryan, G.J. (2005) A single domestication for potato based on multilocus amplified fragment length polymorphism genotyping. *Proceedings of the National Academy of Sciences USA* 102: 14694–9.
307. Hawkes, J.G. (1990) *The potato: Evolution, biodiversity and genetic resource*. Belhaven Press, London.
308. Rios, D., Ghislain, M., Rodríguez, F., and Spooner, D.M. (2007) What is the origin of the European potato? Evidence from Canary Island landraces. *Crop Science* 47: 1271–80.
309. Gerard, J. (1597) *The herball or generall historie of plantes*. John Norton, London, p. 782.
310. Gomez-Alpizar, L,, Carbone, I., and Ristaino, J.B. (2007) An Andean origin of *Phytophthora infestans* inferred from mitochondrial and nuclear gene genealogies. *Proceedings of the National Academy of Sciences USA* 104: 3306–11. Grunwald, N.J., and Flier, W.G. (2005) The biology of *Phytophthora infestans* at its center of origin. *Annual Review of Phytopathology* 43: 171–90.
311. Kinealy, C. (1995) *This great calamity: The Irish Famine 1845–52*. Gill & Macmillan, Dublin.

312. Hawkes, J.G. (1990) *The potato: Evolution, biodiversity and genetic resource*. Belhaven Press, London.
313. Rios, D., Ghislain, M., Rodríguez, F., and Spooner, D.M. (2007) What is the origin of the European potato? Evidence from Canary Island landraces. *Crop Science* 47: 1271–80.
314. Ames, M., and Spooner, D.M. (2008) DNA from herbarium specimens settles a controversy about origins of the European potato. *American Journal of Botany* 95: 252–7.
315. George and Ira Gershwin, 'Let's call the whole thing off' (1937).
316. Gentilcore, D. (2010). *Pomodoro! A history of the tomato in Italy*. Columbia University Press, New York.
317. Peralta, I.E., Spooner, D.M., and Knapp, S. (2008) *Taxonomy of wild tomatoes and their relatives (Solanum sect. Lycopersicoides, sect. Juglandifolia, sect. Lycopersicon; Solanaceae)*. American Society of Plant Taxonomists, Ann Arbor.
318. Peralta, I.E., Spooner, D.M., and Knapp, S. (2008) Taxonomy of wild tomatoes and their relatives (*Solanum* sect. *Lycopersicoides*, sect. *Juglandifolia*, sect. *Lycopersicon*; Solanaceae). American Society of Plant Taxonomists, Ann Arbor.
319. Blanca, J., Cañizares, J., Cordero, L., Pascual, L., Diez, M.J., and Nuez, F. (2012) Variation revealed by SNP genotyping and morphology provides insight into the origin of the tomato. *PLoS ONE* 7: e48198.
320. Gentilcore, D. (2010). *Pomodoro! A history of the tomato in Italy*. Columbia University Press, New York.
321. Gerard, J. (1597) *The herball or generall historie of plantes*. John Norton, London, pp. 275–6.
322. Parkinson, J. (1629) *Paradisi in sole*. Printed by Humfrey Lownes & Robert Young, London, pp. 379–80.
323. Bobart, J. (1648) *Catalogus Plantarium Hortii Medici Oxoniensis*. University of Oxford, Oxford. Stephens, P., and Brown, W. (1658) *Catalogus Horti Botanici Oxoniensis*. William Hall, Oxford.
324. Gerard, J. (1597) *The herball or generall historie of plantes*. John Norton, London, pp. 275–6.
325. Parkinson, J. (1629) *Paradisi in sole*. Printed by Humfrey Lownes & Robert Young, London, pp. 379–80.
326. Vilmorin-Andrieux, M. (1885) *The vegetable garden*. John Murray, London.
327. Monforte, A.J., Diaz, A.I., Caño-Delgado, A. van der Knaap, E. (2014) The genetic basis of fruit morphology in horticultural crops: lessons from tomato and melon. *Journal of Experimental Botany* 65: 4625–37.
328. Martineau, B. (2001) *First fruit: The creation of the Flavr Savr tomato and the birth of biotech food*. McGraw-Hill Education, London.
329. Butelli, E., Titta, L., Giorgio, M., Mock, H.-P., Matros, A., Peterek, S., Schijlen, E., Hall, R., Bovy, A., Luo, J., and Martin, C. (2008) Induced anthocyanin biosynthesis in tomato results in purple fruit with increased antioxidant and dietary, health-protecting properties. *Nature Biotechnology* 26: 1301–8. Gonzali, S., Mazzucato, A., and Perata, P. (2009) Purple as a tomato: towards high anthocyanin tomatoes. *Trends in Plant Science* 14: 237–41. Zhang, Y., Butelli, E., de Stefano, R., Schoonbeek, H.J., Magusin, A., Pagliarani, C., Wellner, N., Hill, L., Orzaez, D., Granell, A., Jones, J.D.G., and Martin, C. (2013) Anthocyanins double the shelf

life of tomatoes by delaying over-ripening and reducing susceptibility to gray mold. *Current Biology* 23: 1094–100.
330. Wroth, W.W., and Kell, P.E. (2004) Salter, James (d. *c.* 1728). *Oxford Dictionary of National Biography*, Oxford University Press, Oxford. www.oxforddnb.com/view/article/24569 (accessed 4 January 2024).
331. Anonymous (*c.* 1785) *A catalogue of the rarities, to be seen at Don Saltero's coffee-house in Chelsea*, 38th edn. London.
332. Lillywhite, B. (1963) *London coffee houses*. George Allen & Unwin, London.
333. Ellis, J. (1774) *An historical account of coffee*. Edward and Charles Dilly, London.
334. Bradley, R. (1718) *New improvements of planting and gardening, both philosophical and practical; explaining the motion of the sapp. and generation of plants*. W. Mears, London, p. 235.
335. Moseley, B. (1792) *A treatise concerning the properties and effects of coffee*. J. Sewell, London, p. v.
336. Saint-Arroman, A. (1852) *Coffee, tea and chocolate: Their influence upon the health, the intellect and the moral nature of man*. Crissy & Marklet, Philadelphia PA, p. 30.
337. Knight, F.W. (2000) The Haitian Revolution. *American Historical Review* 105: 103–15.
338. Piddington, H. (1839) *A letter to the European soldiers in India, on the substitution of coffee for spirituous liquors*. Englishman Press, London. Saint-Arroman, A. (1852) *Coffee, tea and chocolate: Their influence upon the health, the intellect and the moral nature of man*. Crissy & Marklet, Philadelphia. Weinberg, B.A., and Bealer, B.K. (2002) *The world of caffeine: The science and culture of the world's most popular drug*. Routledge, London.
339. Moseley, B. (1792) *A treatise concerning the properties and effects of coffee*. J. Sewell, London, p. 2. Simmonds, P.L. (1850) *Coffee as it us, and as it ought to be*. Effingham Wilson, London, p. 5.
340. Anonymous (1832) *A treatise in the nature and properties of kalopino, the new invented substitute for tea and coffee*. J.T. Norris, London.
341. Gross, M. (2014) Coffee and chocolate in danger. *Current Biology* 24: R503–R506.
342. Mazoyer, M., and Roudart, L. (2006) *A history of world agriculture from the Neolithic Age to the current crisis*. Earthscan, London.
343. Gerard, J. (1597) *The herball or generall historie of plantes*. John Norton, London, p. 77.
344. Sauer, C.O. (1960) Maize in Europe. Akten des 34. Internationalen Amerikanisten Kongresses, Vienna, pp. 778–86.
345. De Candolle, A. (1837) Histoire. Histoire naturelle, agricole et économique du maïs, par M. Matthieu Bonafous. *Annales de l'agriculture française* 19: 5–27. Bonafous, M. (1833) *Traité du maïs, ou histoire naturelle et agricole de cette céréale*. Mme Huzard, Paris. 2nd edn published in folio as Bonafous, M. (1836) *Histoire naturelle, agricole et économique du maïs*. Mme Huzard, Paris.
346. Kunz, K., and Sigurdsson, G. (2008) *The Vinland sagas*. Penguin, London, pp. 35, 44, 56 n11. Sturtevant, E.L. (1885) Indian corn and the Indian. *The American Naturalist* 19: 225–34.
347. Benz, B.F. (2001) Archaeological evidence of teosinte domestication of Guilá Naquitz, Oaxaca. *Proceedings of the National Academy of Sciences USA* 98: 2104–6. Piperno, D.R., and Flannery, K.V. (2001) The earliest archaeological maize (*Zea*

mays L.) from Highland Mexico: new accelerator mass spectrometry dates and their implication. *Proceedings of the National Academy of Sciences USA 98*: 2101–3.

348. Beadle, G.W. (1939) Teosinte and the origin of maize. *Journal of Heredity* 30: 245–7. Jaenicke-Després, V., Buckler, E.S., Smith, B.D., Gilbert, M.T.P., Cooper, A., Doebley, J., and Pääbo, S. (2003) Early allelic selection in maize as revealed by ancient DNA. *Science* 302: 1206–8. Doebley, J. (2004) The genetics of maize evolution. *Annual Review of Genetics* 38: 37–59.
349. Ralph, J. (1758) *The case of authors by profession or trade stated*. Printed by R. Griffiths, London, p. 41.
350. MacNutt, F.A. (1912) *De Orbe Novo: the eight decades of Peter Martyr d'Anghera*, Vol. 1. G.P. Putnam & Sons, London, p. 262.
351. Do Rosário, A. (1702) *Frutas do Brasil numa nova, e ascetica Monarchia, consagrada à Santissima Senhora do Rosario*. António Galrão, Lisbon. Biron, B.R.R. (2009) *Frutas do Brasil*: Uma alegoria do Novo Mundo. *Revista do Núcleo de Estudos de Literatura Portuguesa e Africana da UFF* 2: 47–57.
352. Whatley, J. (1992) *Jean de Léry: History of a voyage to the land of Brazil, otherwise called America*. University of California Press, Berkeley, p. 108.
353. Whatley, J. (1992) *Jean de Léry: History of a voyage to the land of Brazil, otherwise called America*. University of California Press, Berkeley, p. 108. Piso, W., and Marggraf, G. (1648) *Historia naturalis Brasiliae*. Franciscum Hackium, Amstelodami, apud Lud. Elzevirium, Lugdunum Batavorum.
354. Boym, M. (1656) *Flora Sinensis, fructus floresque humillime porrigens*. Typis Matthaei Rictiji, Viennae, tab. G.
355. Ligon, R. (1657) *True and exact history of Barbados*. Humphrey Moseley, London.
356. Parkinson, J. (1640) *Theatrum Botanicum: The theatre of plantes*. Tho. Cotes, London, pp. 1626–7.
357. De Beer, E.S. (1955) *The diary of John Evelyn*, Vol. 3. Clarendon Press, Oxford, p. 293.
358. Bradley, R. (1721) *A general treatise of husbandry and gardening*, Vol. 1. Printed for T. Woodward & J. Peele, London, pp. 206–20.
359. Musgrave, T. (2009) *The head gardeners: Forgotten heroes of horticulture*. Aurum Press, London.
360. Brown, J. (2000) *The pursuit of paradise: A social history of gardens and gardening*. HarperCollins, London.
361. Pearson, R.A., Lhoste, P., Saatamoinen, M., and Martin-Rosset, W. (2003) *Working animals in agriculture and transport: A collection of some current research and development observations*. Wageningen Academic Publishers, Wageningen.
362. Flint, C.L. (1859) *Grasses and forage plants: A practical treatise*. W.F. Gill, Boston MA, p. 91.
363. Hitchcock, A.S. (1935) *Manual of the grasses of the United States*. United States Government Printing Office, Washington DC.
364. Jefferson, R.G. (2005) The conservation management of upland hay meadows in Britain: A review. *Grass and Forage Science* 60: 322–31.
365. Berridge, V. (2005) *The Big Smoke: Fifty years after the 1952 London smog*. University of London, London. Bell, M.L., Devra, L., and Davis, T.F. (2004) A retrospective assessment of mortality from the London Smog episode of 1952: The role of influenza and pollution. *Environmental Health Perspectives* 112: 6–8.

366. Owen, D. (2004) *Copies in seconds: How a lone inventor and an unknown company created the biggest communication breakthrough since Gutenberg*. Simon & Schuster, New York.
367. McHenry, G. (1863) *The cotton trade: Its bearing upon the prosperity of Great Britain and commerce of the American Republics, considered in connection with the system of Negro slavery in the Confederate State*. Saunders, Otley & Co., London.
368. Lee, H. (1887) *The vegetable lamb of Tartary: A curious fable of the cotton plant*. Sampson Low, Marston, Searle & Rivington, London. Appleby, J.H. (1997) The Royal Society and the Tartar Lamb. *Notes and Records of the Royal Society of London* 51: 23–34.
369. Watt, G. (1907) *The wild and cultivated cotton plants of the world: A revision of the genus* Gossypium *framed primarily with the object of aiding planters and investigators who may contemplate the systematic improvement of the cotton staple*. Longmans, Green, & Co., London.
370. Hort, A. (1916) *Theophrastus: Enquiry into plants*, Vol. 1. William Heinemann, London, chs 4.4 and 4.7.
371. Thomson, J. (1849) On the mummy cloth of Egypt. *The Classical Museum* 6: 152–63.
372. Fryxell, P.A. (1979) *The natural history of the cotton tribe (Malvaceae, tribe Gossypeae)*. A&M University Press, College Station TX and London.
373. Huckell, L.W. (1993) Plant remains from the Pinaleño Cotton Cache, Arizona. *Kiva, the Journal of Southwest Anthropology and History* 59: 147–203. Moulherat, C., Tengberg, M., Haquet, J.F., and Mille, B. (2002) First evidence of cotton at Neolithic Mehrgarh, Pakistan: Analysis of mineralized fibres from a copper bead. *Journal of Archaeological Science* 29: 1393–1401. Wendel, J.F. (1989) New World tetraploid cottons contain Old World cytoplasm. *Proceedings of the National Academy of Sciences USA* 86: 4132–6.
374. Bowman, F.H. (1908) *The structure of the cotton fibre in its relation to technical applications*. Macmillan, London. Flint, E.A. (1950) The structure and development of the cotton fibre. *Biological Reviews* 25: 414–34.
375. Fryxell, P.A. (1979) *The natural history of the cotton tribe (Malvaceae, tribe Gossypeae)*. A&M University Press, College Station TX and London.
376. Drury, H. (1873) *The useful plants of India*. William H. Allen, London, p. 230.
377. McHenry, G. (1863) *The cotton trade: Its bearing upon the prosperity of Great Britain and commerce of the American Republics, considered in connection with the system of Negro slavery in the Confederate State*. Saunders, Otley & Co., London. Hobhouse, H. (1999) *Seeds of change*. Papermac, London.
378. Watts, J. (1866) *The facts of the cotton famine*. Simpkin, Marshall, & Co., London. Arnold, A. (1864) *The history of the cotton famine from the fall of Sumter to the passing of the Public Works Act*. Saunders, Otley, & Co., London.
379. Davis, M. (2000) *Late Victorian Holocausts: El Niño famines and the making of the Third World*. Verso, London.
380. Perrin, L.M. (2001) Resisting reproduction: Reconsidering slave contraception in the Old South. *Journal of American Studies* 35: 255–74.
381. Waltham, T., and Sholji, I. 2001. The demise of the Aral Sea – an environmental disaster. *Geology Today* 17: 218–28. Wang, Z., Lin, H., Huang, J., Hu, R., Rozelle, S., and Pray, C. 2009. Bt cotton in China: Are secondary insect infestations

offsetting the benefits in farmer fields? *Agricultural Sciences in China* 8: 83–90. Lu, W.K., Wu, K., Jiang, Y., Guo, Y., and Desneux, N. (2012) Widespread adoption of Bt cotton and insecticide decrease promotes biocontrol services. *Nature* 487: 362–5.

382. 'Os Escravos saõ as mãos, & os pés do Senhor do Engenho; porque sem elles no Brasil naõ he possivel fazer, conservar, & aumentar Fazenda, nem ter Engenho'. Antonil, A.J. (1711) Cultura, e opulencia do Brasil por suas drogas, e minas. Officina Real Deslandesiana, Lisbon, p. 22.
383. Moseley, B. (1799) *A treatise on sugar*. Printed for G.G., and J. Robinson, London.
384. Gerard, J. (1597) *The herball or generall historie of plantes*. John Norton, London, p. 35.
385. Moseley, B. (1799) *A treatise on sugar*. Printed for G.G., and J. Robinson, London, p. 157.
386. Voltaire (1759) *Candide, ou L'optimisme, traduit de l'Allemand de Mr. le Docteur Ralph*. Paris, p. 127.
387. Moseley, B. (1799) *A treatise on* sugar. Printed for G.G., and J. Robinson, London, p. 140.
388. Treloar, W.P. (1884) *The prince of palms*. Sampson Low, Marston, Searle & Rivington, London, p. 5.
389. Treloar, W.P. (1884) *The prince of palms*. Sampson Low, Marston, Searle & Rivington, London.
390. Purseglove, J.W. (1972) *Tropical crops: Monocotyldons*. Longman: London.
391. Halliwell, J.O. (1839) *The voyage and travaile of Sir John Maundevile, Kt*. Edward Lumley, London, p. 265.
392. Gerard, J. (1597) *The herball or generall historie of plantes*. John Norton, London, pp. 1337–9. Fritz, R. (1983). *Die Gefässe aus Kokosnuss in Mitteleuropa 1250–1800*. Philipp. von Zabern, Mainz am Rhein.
393. Harries, H.C. (2004) Fun made the fair coconut shy. *Palms* 48: 77–82.
394. Edmondson, C.H. (1941) Viability of coconut after floating in sea. *Occasional Papers of the Bernice Pauahi Bishop Museum of Polynesian Ethnology and Natural History, Hawaii* 16: 293–304.
395. Hartwig, G. (1863) *The tropical world: A popular scientific account of the natural history of the animal and vegetable kingdoms in the equatorial regions*. Longman, Green, Longman, Roberts, & Green, London, pp. 128–9.
396. Clement, C.R., Zizumbo-Villarreal, D., Brown, C.H., Ward, R.G., Alves-Pereira, A., and Harries, H.C. (2013) Coconuts in the Americas. *Botanical Review* 79: 342–70. Gunn, B.F., Baudouin, L., Olsen, K.M. (2011) Independent origins of cultivated coconut (*Cocos nucifera* L.) in the Old World tropics. *PLoS ONE* 6: e21143.
397. Clement, C.R., Zizumbo-Villarreal, D., Brown, C.H., Ward, R.G., Alves-Pereira, A., and Harries, H.C. (2013) Coconuts in the Americas. *Botanical Review* 79: 342–70.
398. De Candolle, A. (1884) *Origin of cultivated plants*. Kegan Paul, Trench & Co., London. Purseglove, J.W. (1972) *Tropical crops: Monocotyldons*. Longman, London. Gunn, B.F., Baudouin, L., and Olsen, K.M. (2011) Independent origins of cultivated coconut (*Cocos nucifera* L.) in the Old World tropics. *PLoS ONE* 6: e21143.
399. Cook, O.F. (1901) *The origin and distribution of the cocoa palm*. Government

Printing Office, Washington DC. Harries, H.C. (1992) Biogeography of the coconut *Cocos nucifera* L. *Principes* 36: 155–62. Clement, C.R., Zizumbo-Villarreal, D., Brown, C.H., Ward, R.G., Alves-Pereira, A., and Harries, H.C. (2013) Coconuts in the Americas. *Botanical Review* 79: 342–70. Baker, W.J., and Couvreur, T.L.P. (2013) Global biogeography and diversification of palms sheds light on the evolution of tropical lineages. I. Historical biogeography. *Journal of Biogeography* 40: 274–85. Meerow, A.W., Noblick, L., Borrone, J.W., Couvreur, T.L.P., Mauro-Herrera, M., Hahn, W.J., Kuhn, D.N., Nakamura, K., Oleas, N.H., and Schnell, R.J. (2009) Phylogenetic analysis of seven WRKY genes across the palm subtribe Attaleinae (Arecaceae) identifies *Syagrus* as sister group of the coconut. *PLoS ONE* 4: e7353.

400. Harries, H.C., and Clement, C.R. (2014) Long-distance dispersal of the coconut palm by migration within the coral atoll ecosystem. *Annals of Botany* 113: 565–70.
401. Harries, H.C. (2004) Fun made the fair coconut shy. *Palms* 48: 77–82. Jackson, J.R. (1890) *Commercial botany of the nineteenth century.* Cassell & Co., London.
402. Treloar, W.P. (1884) *The prince of palms.* Sampson Low, Marston, Searle & Rivington, London.
403. Anonymous (1912) *The cult of the coconut: A popular exposition of the coconut and oil palm industries.* Curtis Gardner, London.
404. Purseglove, J.W. (1972) *Tropical crops. Monocotyledons.* Longman: London.
405. Carney, J.A. (2001) *Black rice: The African origins of rice cultivation in the Americas.* Harvard University Press, Cambridge MA. Carney, J.A., and Rosomoff, R.N. (2009) *In the shadow of slavery: Africa's botanical legacy in the Atlantic World.* University of California Press, Berkeley CA.
406. Qian, H., Zhu, X., Huang, S., Linquist, B., Kuzyakov, Y., Wassmann, R., Minamikawa, K., Martinez-Eixarch, M., Yan, X., Zhou, F., Sander, B.O., Zhang, W., Shang, Z., Zou, J., Zheng, X., Li, G., Liu, Z., Wang, S., Ding, Y., van Groenigen, K.J., and Jiang, Y. (2023) Greenhouse gas emissions and mitigation in rice agriculture. *Nature Reviews Earth & Environment* 4: 716–32.
407. Sweeney, M., and McCouch, S. (2007) The complex history of the domestication of rice. *Annals of Botany* 100: 951–7. Fuller, D.Q., Sato, Y.-I., Castillo, C., Qin, L., Weisskopf, A.R., Kingwell-Banham, E.J., Song, J., Ahn, S.-M., and van Etten, J. (2010) Consilience of genetics and archaeobotany in the entangled history of rice. *Archaeological and Anthropological Sciences* 2: 115–31. Molina, J., Sikora, M., Garud, N., Flowers, J.M., Rubinstein, S., Reynolds, A., Huang, P., Jackson, S., Schaal, B.A., Bustamante, C.D., Boyko, A.R., and Purugganan, M.D. (2011) Molecular evidence for a single evolutionary origin of domesticated rice. *Proceedings of the National Academy of Sciences USA* 108: 8351–6.
408. Li, Z.M., Zheng, X.M., and Ge, S. (2011) Genetic diversity and domestication history of African rice (*Oryza glaberrima*) as inferred from multiple gene sequences. *Theoretical and Applied Genetics* 123: 21–31. Portères, R. (1970) *Primary cradles of agriculture in the African continent.* Cambridge University Press, Cambridge. Richards, P. (1996) *Redefining nature: Ecology, culture and domestication.* Oxford University Press, Oxford.
409. Linares, O.F. (2002) African rice (*Oryza glaberrima*): History and future potential. *Proceedings of the National Academy of Sciences USA* 99: 16360–65.

410. Ermakova,M., Danila, F.R., Furbank, R.T., and von Caemmerer, S. (2020) On the road to C_4 rice: Advances and perspectives. *The Plant Journal* 101: 940–50.
411. Beyer, P. (2010) Golden rice and 'golden' crops for human nutrition. *New Biotechnology* 27: 478–81. Salim Al-Babili, S., and Beyer, P. (2005) Golden Rice – five years on the road – five years to go? *Trends in Plant Science* 10: 565–73. Eisenstein, M. (2014) Biotechnology: against the grain. *Nature* 514: S55–S57. de Steur, H., Stein, A.J., and Demont, M. (2022) From Golden Rice to Golden Diets: How to turn its recent approval into practice. *Global Food Security* 32: 100596.
412. Koerner L (1999) *Linnaeus: Nature and nation*. Harvard University Press, Cambridge MA.
413. Hamilton, H. (1948) *England: A history of the homeland*. Norton, New York.
414. Smith, J. (1848) *Essays on the cultivation of the tea plant, in the United States of America*. W.E. Dean, New York, p. 3.
415. Gardner, G. (1846) *Travels in the interior of Brazil, principally through the Northern Provinces and the gold and diamond districts, during the years 1836–1841*. Reeve Brothers, London, p. 35.
416. Fortune, R. (1847) *Three years' wandering in the northern provinces of China, a visit to the tea, silk, and cotton countries*. John Murray, London. Fortune, R. (1852) *A journey to the tea countries of China*. John Murray, London.
417. Fortune, R. (1847) *Three years' wandering in the northern provinces of China, a visit to the tea, silk, and cotton countries*. John Murray, London.
418. Harbowy, M.E., and Balentine, D.A. (1997) Tea chemistry. *Critical Reviews in Plant Sciences* 16: 415–80.
419. Brockway, L.H. (1979) *Science and colonial expansion: The role of the British Royal Botanic Gardens*. Yale University Press, New Haven CT.
420. Ward, N.B. (1852) *On the growth of plants in closely glazed cases*. John von Voorst, London. Keogh, L. (2020) *The Wardian case: How a simple box moved plants and changed the world*. University of Chicago Press, Chicago IL.
421. Blatchley, W.S. (1912) *The Indiana weed book*. Nature Publishing Co., Indianapolis. Wodehouse, R.P. (1960) Weed. *Encyclopaedia Britannica* 23: 477.
422. Pelser, P.B., van den Hof, K., Gravendeel, B., and van der Meijden, R. (2004) The systematic value of morphological characters in *Senecio* sect. Jacobaea (Asteraceae). *Systematic Botany* 29: 790–805. Pelser, P.B., Nordenstam, B., Kadereit, J.W., and Watson, L.E. (2007) An ITS phylogeny of tribe Senecioneae (Asteraceae) and a new delimitation of *Senecio* L. *Taxon* 56: 1077–104.
423. Harper, J.L., and Wood, W.A. (1957) Biological flora of the British Isles. *Senecio jacobaea* L. *Journal of Ecology* 45: 617–37.
424. Cameron, E. (1935) A study of the natural control of ragwort (*Senecio jacobaea* L.). *Journal of Ecology* 23: 265–322. Schmidl, L. (1972) Biology and control of ragwort, *Senecio jacobaea* L., in Victoria, Australia. *Weed Research* 12: 37–45.
425. Rizk, A.F.M. (1991) The pyrrolizidine alkaloids: Plant sources and properties. In Rizk, A.F.M. (ed.) *Naturally occurring pyrrolizidine alkaloids*. CRC Press, Boca Raton FL, pp. 1–89.
426. Peterson, J.E., and Culvenor, C.C.J. (1983) Hepatotoxic pyrrolizidine alkaloids. In Keeler, R.F., and Tu, A.T. (eds) *Handbook of natural toxins*, Volume 1: *Plant and fungal toxins*. Marcel Dekker, New York, pp. 637–71.
427. Noonan, P.J. (2001) Common plant poisoning. *Irish Veterinary Journal* 54: 342–4.

428. Pemberton, R.W., and Turner, C.E. (1990). Biological control of *Senecio jacobaea* in northern California, an enduring success. *Entomophaga* 35: 71–7.
429. Grieve, M. (1931) *A modern herbal*, Vol. 2. Jonathan Cape, London.
430. Gerard, J. (1597) *The Herball or Generall Historie of Plantes*. John Norton, London, pp. 1331–3.
431. Denham, T.P., Haberle, S.G., Lentfer, C., Fullagar, R., Field, J., Therin, M., Porch, N., and Winsborough, B. (2003). Origins of agriculture at Kuk Swamp in the highlands of New Guinea. *Science* 5630: 189–93.
432. Simmonds, N.W., and Shepherd, K. (1955) Taxonomy and origins of cultivated bananas. *Botanical Journal of the Linnean Society* 55: 302–12.
433. Watson, A. (1983) *Agricultural innovation in the early Islamic world*. Cambridge University Press, Cambridge.
434. Johnson, T. (1636) *The Herball or Generall Historie of Plantes. Gathered by John Gerarde of London*. Adam Islip Ioice Norton & Richard Whitakers, London, p. 1515.
435. Freer, S. (2007) *Musa cliffortiana: Clifford's banana plant*. Reprint and translation of the original edition (Leiden 1736). A.R.G. Gantner Verlag, Ruggell. Watson, A. (1983) *Agricultural innovation in the early Islamic world*. Cambridge University Press, Cambridge. Johnson, T. (1636) *The Herball or Generall Historie of Plantes. Gathered by John Gerarde of London*. Printed by Adam Islip, Ioice Norton & Richard Whitakers, London, pp. 1331–3.
436. Purseglove, J.W. (1972) *Tropical crops: Monocotyledons*. Longman, London.
437. Moore, N.Y., Bentley, S., Pegg, K.G., and Jones, D. R. (1995) *Musa disease fact sheet No 5. Fusarium wilt of banana*. INIBAP, Montpellier.
438. Marín, D.H., Romero, R.A., Guzmán, M., and Sutton, T.B. (2003) Black sigatoka: An increasing threat to banana cultivation. *Plant Disease* 87: 208–22.
439. Hwang, S.-C., and Ko, W.-H. (2004) Cavendish banana cultivars resistant to *Fusarium* Wilt acquired through somaclonal variation in Taiwan. *Plant Disease* 6: 580–88. Robert, S., Ravigne, V., Zapater, M.F., Abadie, C., and Carlier, J. (2012) Contrasting introduction scenarios among continents in the worldwide invasion of the banana fungal pathogen *Mycosphaerella fijiensis*. *Molecular Ecology* 21: 1098–114.
440. d'Hont, A., Denoeud, F., Aury, J.-M., Baurens, F.-C., Carreel, F., Garsmeur, O., Noel, B., Bocs, S., Droc, G., Rouard, M., Da Silva, C., Jabbari, K., Cardi, C., Poulain, J., Souquet, M., Labadie, K., Jourda, C., Lengelle, J., Rodier-Goud, M., Alberti, A., Bernard, M., Correa, M., Ayyampalayam, S., McKain, M.R., Leebens-Mack, J., Burgess, D., Freeling, M., Mbeguie-A-Mbeguie, D., Chabannes, M., Wicker, T., Panaud, O., Barbosa, J., Hribova, E., Heslop-Harrison, P., Habas, R., Rivallan, R., Francois, P., Poiron, C., Kilian, A., Burthia, D., Jenny, C., Bakry, F., Brown, S., Guignon, V., Kema, G., Dita, M., Waalwijk, C., Joseph, S., Dievart, A., Jaillon, O., Leclercq, J., Argout, X., Lyons, E., Almeida, A., Jeridi, M., Dolezel, J., Roux, N., Risterucci, A.-M., Weissenbach, J., Ruiz, M., Glaszmann, J.-C., Quetier, F., Yahiaoui, N., and Wincker, P. (2012) The banana (*Musa acuminata*) genome and the evolution of monocotyledonous plants. *Nature* 488: 213–17.
441. Whittington, E.M. (2001) *The sport of life and death: The Mesoamerican ballgame*. Thames & Hudson, New York.
442. Filloy Nadal, L. (2001) Rubber and rubber balls in Mesoamerica. In Whittington,

E.M. (ed.) *The sport of life and death: The Mesoamerican ballgame*. Thames & Hudson, New York, pp. 20–31.
443. Dean, W. (1987) *Brazil and the struggle for rubber: A study in environmental history*. Cambridge University Press, Cambridge.
444. Cowell, A. (1990) *The decade of destruction*. Hodder & Stoughton, London, pp. 169–200.
445. Moerman, D.E. (1998) *Native American ethnobotany*. Timber Press, Portland OR, pp. 257–8. Heiser, C.B. (1951) The sunflower among the North American Indians. *Proceedings of the American Philosophical Society* 95: 432–48.
446. Gerard, J. (1597) *The Herball or general historie of plantes. Gathered by John Gerarde of London Master in Chirurgerie*. John Norton, London, p. 612.
447. Parkinson, J. (1629) *Paradisi in sole*. Printed by Humfrey Lownes & Robert Young, London, pp. 295–6.
448. Knapp, S. (2014) Why is a raven like a writing desk? Origins of the sunflower that is neither an artichoke nor from Jerusalem. *New Phytologist* 201: 710–11.
449. Gerard, J. (1633) *The herball or Generall historie of plantes*. Printed by Adam Islip, Ioice Norton and Richard Whitakers, London, p. 754.
450. Gerard, J. (1597) *The Herball or general historie of plantes. Gathered by John Gerarde of London Master in Chirurgerie*. John Norton, London, p. 613.
451. Adam, J.A. (2009) *A mathematical nature walk*. Princeton University Press, Princeton NJ, pp. 31–42. Swinton, J., Ochu, E. and MSI Turing's Sunflower Consortium (2016) Novel Fibonacci and non-Fibonacci structure in the sunflower: Results of a citizen science experiment. *Royal Society Open Science* 3: 160091.
452. Mathai, A.M., and Davis, T.A. (1974) Constructing the sunflower head. *Mathematical Biosciences* 20: 117–33. Vogel, H. (1979) A better way to construct the sunflower head. *Mathematical Biosciences* 44: 179–89.
453. Heiser, C.B. (1998) The domesticated sunflower in old Mexico? *Genetic Resources and Crop Evolution* 45: 447–9. Lentz, D.L., Pohl, M.D., Alvarado, J.L., Tarighat, S., and Bye, R. (2008) Sunflower (*Helianthus annuus* L.) as a pre-Columbian domesticate in Mexico. *Proceedings of the National Academy of Sciences USA* 105: 6232–7. Lentz, D.L., Pohl, M.D., Pope, K.O., and Wyatt, A.R. (2001) Prehistoric sunflower (*Helianthus annuus* L.) domestication in Mexico. *Economic Botany* 55: 370–76. Heiser, C.B. (2008) Sunflowers among Aztecs? *International Journal of Plant Sciences* 169: 980. Lentz, D.L. (2008) Reply to Heiser. *International Journal of Plant Sciences* 169: 980. Heiser, C.B. (2008) The sunflower (*Helianthus annuus*) in Mexico: Further evidence for a North American domestication. *Genetic Resources and Crop Evolution* 55: 9–13.
454. Putt, E.D. (1997) Early history of sunflower. In Schneiter, A.A. (ed.) *Sunflower technology and production*. American Society of Agronomy, Crop Science Society of America and Soil Science Society of America, Madison WI.
455. Watkins, C. (2006) Operation Oilseed. *Sunflower Magazine*. www.sunflowernsa.com/magazine/details.asp?ID=420&Cat=1 (accessed 4 January 2024).
456. Burkill, H.M. 1997 *The useful plants of West Tropical Africa*, Volume 4: *Families M–R*. Royal Botanic Gardens, Kew, pp. 354–69.
457. Sir Hans Sloane's *Catalogus Plantarum* (1696) describes oil palm in Jamaica. William Dampier, in his *A voyage to New Holland, &c. In the Year, 1699* (1703: p. 71), may refer to the African oil palm or perhaps a Brazilian oil-producing palm.

Loudon, J.C. (1850) *Loudon's Hortus Britannicus*. Printed for Longman, Brown, Green, and Longmans, London, p. 399.
458. Purseglove, J.W. (1972) *Tropical crops: Monocotyledons*. Longman, London.
459. Secretariat of the Convention on Biological Diversity (2010) *Global Biodiversity Outlook* 3. Montréal, p. 5.
460. Meijaard, E., Garcia-Ulloa, J., Sheil, D., Wich, S.A., Carlson, K.M., Juffe-Bignoli, D., and Brooks, T.M. (2018). *Oil palm and biodiversity: A situation analysis by the IUCN Oil Palm Task Force*. IUCN, Gland, Switzerland.
461. Singh, R., Ong-Abdullah, M., Low, E.-T.L., Manaf, M.A.A., Rosli, R., Nookiah, R., Ooi, L.C.-L., Ooi, S.-E., Chan, K.-L., Halim, M.A., Azizi, N., Nagappan, J., Bacher, B., Lakey, N., Smith, S.W., He, D., Hogan, M., Budiman, M.A., Lee, E.K., DeSalle, R., Kudrna, D., Goicoechea, J.L., Wing, R.A., Wilson, R.K., Fulton, R.S., Ordway, J.M., Martienssen, R.A., and Sambanthamurthi, R. (2013) Oil palm genome sequence reveals divergence of interfertile species in Old and New worlds. *Nature* 500: 335–9.
462. Hymowitz, T., and Harlan, J.R. (1983) Introduction of soybean to North America by Samuel Bowen in 1765. *Economic Botany* 37: 371–9.
463. Benson Ford Research Center (2014) www.thehenryford.org/research/soybeancar.aspx (accessed 4 January 2024).
464. Kaempfer, E. (1712) *Amoenitatum exoticarum politico-physico-medicarum fasciculi V*. Lemgoviae: Typis & Impensis Henrici Wilhelmi Meyeri, Aulae Lippiacae Typographi, pp. 837–40.
465. Hymowitz, T. (1970) On the domestication of soybean. *Economic Botany* 24: 408–21. Stark, M.T. (2005). *Archaeology of Asia*. Wiley–Blackwell, Hoboken NJ.
466. Vranken, L., Avermaete, T., Petalios, D., and Mathijs, E. (2014) Curbing global meat consumption: Emerging evidence of a second nutrition transition. *Environmental Science & Policy* 39: 95–106.
467. National Agricultural Statistics Service (2014) www.nass.usda.gov/index.asp (accessed 2 January 2044).
468. Mittermeier, R.A., Gil, P.R., Hoffman, M., Pilgrim, J., Brooks, T., Mittermeier, C.G., Lamoreux, J., and da Fonseca, G.A.B. (2005) *Hotspots revisited: Earth's biologically richest and most endangered terrestrial ecoregions*. University of Chicago Press, Chicago IL.
469. WWF (2011) *Soya and the cerrado: Brazil's forgotten jewel*. WWF-UK Report, Cambridge. Instituto Nacional de Pesquisas Espaciais (2023) Divulgação dos dados PRODES Cerrado 2023. www.gov.br/inpe/pt-br/assuntos/ultimas-noticias/a-area-de-vegetacao-nativa-suprimida-no-bioma-cerrado-no-ano-de-2023-foi-de-11-011-70-km2 (accessed 3 January 2024).
470. Ratter, J.A., Ribeiro, J.F., and Bridgewater, S. (1997) The Brazilian cerrado vegetation and threats to its biodiversity. *Annals of Botany* 80: 223–30. Jepson, W. (2005) A disappearing biome? Reconsidering land-cover change in the Brazilian savanna. *Geographical Journal* 171: 99–111.
471. Anonymous (2010) The miracle of the cerrado. *The Economist*, 26 August 2010. See also letters responding to the article in *The Economist*, 9 September 2010.
472. Vince, G. (2014) *Adventures in the anthropocene: A journey to the heart of the planet we made*. Chatto & Windus, London.

473. Thompson, P.A. (1973) Effects of cultivation on the germination character of the corn cockle (*Agrostemma githago* L.). *Annals of Botany* 37: 133–54.
474. Baxter, W. (1837) *British Phaenogamous botany*, Vol. 3. Published by the Author, Oxford, t.175.
475. Gerard, J. (1633) *The herball or Generall historie of plantes. Gathered by Iohn Gerarde of London Master in Chirurgerie very much enlarged and amended by Thomas Iohnson citizen and apothecarye of London*. Printed by Adam Islip, Ioice Norton and Richard Whitakers, London, p. 1087.
476. Graves, G., and Hooker, W.J. (1835) *Curtis's Flora Londinensis*, Vol. 1. Henry G. Bohn, London.
477. Thompson, P.A. (1973) Effects of cultivation on the germination character of the corn cockle (*Agrostemma githago* L.). *Annals of Botany* 37: 133–54. Storkey, J., Meyer, S., Still, K.S., and Leuschner, C. (2012) The impact of agricultural intensification and land-use change on the European arable flora. *Proceedings of the Royal Society B* 279: 1421–9.
478. Storkey, J., Meyer, S., Still, K.S., and Leuschner, C. (2012) The impact of agricultural intensification and land-use change on the European arable flora. *Proceedings of the Royal Society B* 279: 1421–9.
479. Marsh, G.P. (1864) *Man and nature; or, physical geography as modified by human action*. Charles Scribner, New York, p. 50.
480. Egerton, F.N. (2012) *Roots of ecology: Antiquity to Haeckel*. University of California Press, Berkeley CA.
481. Walker, T. (2014) *Plant conservation: Why it matters and how it works*. Timber Press, Portland OR.
482. Walker, T. (2014) *Plant conservation: Why it matters and how it works*. Timber Press, Portland OR.
483. Storkey, J., Meyer, S., Still, K.S., and Leuschner, C. (2012) The impact of agricultural intensification and land-use change on the European arable flora. *Proceedings of the Royal Society B* 279: 1421–9.
484. Thallius, J. (1588) *Sylva Hercynia*. Francofvurti ad Moenum.
485. Holl, F., and Heynhold, G. (1842) *Flora von Sachsen*. Verlag von Justus Raumann, Dresden, p. 538.
486. Mitchell-Olds, T. (2001) *Arabidopsis thaliana* and its wild relatives: a model system for ecology and evolution, *Trends in Ecology & Evolution* 16: 693–700.
487. International Human Genome Sequencing Consortium (2004) Finishing the euchromatic sequence of the human genome. *Nature* 431: 931–45.
488. The *Arabidopsis* Genome Initiative (2000) Analysis of the genome sequence of the flowering plant *Arabidopsis thaliana*. *Nature* 408: 796–815.
489. The Arabidopsis Information Resource (2024) www.arabidopsis.org (accessed 3 January 2024).
490. Coen, E.S., and Meyerowitz, E.M. (1991) The war of the whorls: Genetic interactions controlling flower development. *Nature* 353: 31–7.
491. Meyerowitz, E.M. (2001) Prehistory and history of *Arabidopsis* research. *Plant Physiology* 125: 15–19.

Further reading

Abbott, E. (2010) *Sugar: A bittersweet history*. Duckworth Overlook, London.
Balfour-Paul, J. (2011) *Indigo: Egyptian mummies to blue jeans*. British Museum, London.
Beauman, F. (2005) *The pineapple: King of fruits*. Chatto & Windus, London.
Biancardi, E., Panella, L.W., and Lewellen, R.T. (2012) *Beta maritima: The origin of beets*. Springer, Berlin.
Block, E. (2010) *Garlic and other alliums: The lore and the science*. Royal Society of Chemistry, Cambridge.
Booth, M. (1996) *Opium: A history*. Simon & Schuster, London.
Booth, M. (2003) *Cannabis: A history*. Bantam Books, London.
Brockway, L.H. (1979) *Science and colonial expansion: The role of the British Royal Botanic Gardens*. Yale University Press, New Haven CT.
Brown, D.C. (2011) *King cotton: A cultural, political, and economic history since 1945*. University Press of Mississippi, Jackson MS.
Clarke, R.C., and Merlin, M.D. (2013) *Cannabis: Evolution and ethnobotany*. University of California Press, Berkeley CA.
Coe, S.D., and Coe, M.D. (1996) *The true history of chocolate*. Thames & Hudson, London.
Coles, P. (2019) *Mulberry*. Reaktion Books, London.
Conniff, R (2011) *The species seekers: Heroes, fools, and the mad pursuit of life on earth*. W. W. Norton , New York.
Cook, H.J. (2007) *Matters of exchange: Commerce, medicine, and science in the Dutch Golden Age*. Yale University Press, New Haven CT.
Crosby, A.W. (2003) *The Columbian exchange: Biological and cultural consequences of 1492*. Praeger, Westport CT.
Dash, M. (1999) *Tulipomania: The story of the world's most coveted flower & the extra-ordinary passions it aroused*. Indigo, London.
Diamond, J. (1998) *Guns, germs and steel: A short history of everybody for the last 13,000 years*. Vintage, London.
Dixon, G.R. (2007) *Vegetable brassicas and related crucifers*. CABI, Wallingford.
Dunmire, W.W. (2004) *Gardens of New Spain: How Mediterranean plants and foods changed America*. University of Texas Press, Austin.
Duvall, C (2014) *Cannabis*. Reaktion Books, London.
Endersby, J. (2008) *A guinea pig's history of biology*. Arrow Books, London.
Farrar, L. (2016) *Gardens and gardeners of the Ancient World*. Windgather Press, Oxford.
Fisher, C. (2017) *Tulip*. Reaktion Books, London.

Flannery, M.C. (2023) *In the herbarium: The hidden world of collecting and preserving plants.* Yale University Press, New Haven CT.
Foale, M. (2003) *The coconut odyssey – the bounteous possibilities of the tree of life.* Australian Centre for International Agricultural Research, Canberra.
Food and Agriculture Organization of the United Nations (20132023) *FAOSTAT.* http://faostat3.fao.org.
Francis, J. (2018) *Gardens and gardening in early modern England and Wales 1560–1660.* Yale University Press, New Haven CT.
Freedman, P.H. (2008) *Out of the east: Spices and the medieval imagination.* Yale University Press, New Haven CT.
Freese, B. (2005) *Coal: A human history.* Arrow Books, London.
Gately, I. (2001) *Tobacco: A cultural history of how an exotic plant seduced civilization.* Grove Press, New York.
Gately, I. (2009) *Drink: A cultural history of alcohol.* Gotham Books, New York.
Gledhill, D. (2008) *The names of plants.* Cambridge University Press, Cambridge.
Griebeler, A. (2024) *Botanical icons: Critical practices in the premodern Mediterranean.* University of Chicago Press, Chicago IL.
Hageneder, F. (2013) *Yew.* Reaktion Books, London.
Harris, S.A. (2014) *Grasses.* Reaktion Books, London.
Harris, S.A. (2018) *Sunflowers.* Reaktion Books, London.
Harvey, G. (2001) *The forgiveness of nature: The story of grass.* Vintage, London.
Hobhouse, H. (1999) *Seeds of change.* Papermac, London.
Hochstrasser, J.B. (2007) *Still life and trade in the Dutch Golden Age.* Yale University Press, New Haven CT.
Honigsbaum, M. (2001) *The fever trail: The hunt for the cure for malaria.* Macmillan, London.
Jackson, J. (2008) *The thief at the end of the world: Rubber, power, and the seeds of empire.* Duckworth Overlook, London.
Juniper, B.E., and Mabberley, D.J. (2019) *The extraordinary story of the apple.* Royal Botanic Gardens, Kew.
Kassinger, R (2015) *A garden of marvels: How we discovered that flowers have sex, leaves eat air, and other secrets of plants.* William Morrow, New York.
Kenrick, P., and Davis, P. (2004) *Fossil plants.* Natural History Museum, London.
Keogh, L. (2020) *The Wardian case: How a simple box moved plants and changed the world.* University of Chicago Press, Chicago IL.
Lack, A. (2016) *Poppy.* Reaktion Books, London.
Leigh, G.J. (2004) *The world's greatest fix: A history of nitrogen and agriculture.* Oxford University Press, Oxford.
Mabey, R. (2010). *Weeds: How vagabond plants gatecrashed civilisation and changed the way we think about nature.* Profile Books, London.
Mann, C.C. (2011) *1493: How Europe's discovery of the Americas revolutionized trade, ecology and life on earth.* Granta Books, London.
Markham, C. (1880) *Peruvian bark: A popular account of the introduction of Chinchona cultivation into British India 1860–1880.* John Murray, London.
Mason, L. (2013) *Pine.* Reaktion Books, London.

McNeil, C. (2006). *Chocolate in Mesoamerica: A cultural history of cacao*. University of Florida Press, Gainesville FL.
Milton, G. (1999) *Nathaniel's nutmeg: How one man's courage changed the course of history*. Sceptre, London.
Montgomery, D.R. (2008) *Dirt: The erosion of civilizations*. University of California Press, Berkeley CA.
Morgan, J., and Richards, A. (2002) *The new book of the apples*. Ebury, London.
Morton, A.G. (1981) *History of botanical science: An account of the development of botany from ancient times to the present day*. Academic Press, London.
Murphy, D.J. (2007) *People, plants & genes: The story of crops and humanity*. Oxford University Press, Oxford.
Oakley, H. (2024) *Modern medicines from plants: Botanial histories of some of modern medicine's most important drugs*. Royal College of Physicians, London.
Parkinson, R., and Quirke, S. (1995) *Papyrus*. British Museum Press, London.
Pavord A. (1998) *The tulip*. Bloomsbury, London.
Perdue, R.E., (1958) *Arundo donax* – source of musical reeds and industrial cellulose. *Economic botany* 12: 368–404.
Potter, J. (2010) *The rose: A true history*. Atlantic Books, London.
Reader, J. (2008) *Propitious esculent: The potato in world history*. William Heinemann, London.
Russell, A. and Rahman, E. (2015) *The master plant: Tobacco in lowland South America*. Bloomsbury, London.
Smith, B.D. (1998) *The emergence of agriculture*. Scientific American Library, New York.
Spengler, RN (2019) *Fruit from the sands: The Silk Road origins of the foods we eat*. University of California Press, Oakland CA.
Staller, J.E. (2010) *Maize cobs and cultures: History of Zea mays L*. Springer, Berlin.
Thompson, C.J.S. (1934) *The mystic mandrake*. Rider, London.
Thompson, P. (2010) *Seeds, sex & civilization: How the hidden life of plants has shaped our world*. Thames & Hudson, London.
Walker, K. and Nesbitt, M. (2019) *Just the tonic: A natural history of tonic water*. Kew Publishing, Kew.
Warren, J.M. (2015) *The nature of crops: How we came to eat the plants we do*. CABI, Wallingford.
Weinberg, B.A., and Bealer, B.K. (2002) *The world of caffeine: The science and culture of the world's most popular drug*. Routledge, London.
Wink, M. and van Wyk, B.-E. (2008) *Mind-altering and poisonous plants of the world*. Timber Press, Portland OR.
Young, P. (2012) *Oak*. Reaktion Books, London.
Zohary, D., Hopf, M., and Weiss, E. (2013) *Domestication of plants in the Old World*. Oxford University Press, Oxford.

Picture credits

p. 2, Tea from Elizabeth Blackwell, *A curious herbal*, 1737–39. Sherardian Library of Plant Taxonomy, GC F1-4, plate 352

p. 4, Mandrake and nightshade from Dioscorides, 1240. Bodleian Library, MS. Arab d. 138, fols 120r–119v

p. 9, Pineapple from Weinmann, *Phytanthoza Iconographia*, 1737–44. Bodleian Library, Arch.Nat. hist. G 5, vol. 1, plate 113

p. 10, Chillies from Weinmann, *Phytanthoza Iconographia*, 1737–44. Bodleian Library, Arch.Nat. hist. G 8, vol. 4, plate 930

p. 12, Oak and pine, Tudor pattern book, *c.*1520–30. Bodleian Library, MS. Ashmole 1504, fol. 23v

p. 17, Barley: Johann Wilhelm Weinmann, *Phytanthoza Iconographia*, 1737–44. Bodleian Library, Arch.Nat. hist. G 7, vol. 3, plate 577

p. 23, Mandrake: Johann Wilhelm Weinmann, *Phytanthoza Iconographia*, 1737–44. Bodleian Library, Arch.Nat. hist. G 7, vol. 3, plate 708

p. 28, Beets: Johann Wilhelm Weinmann, *Phytanthoza Iconographia*, 1737–44. Bodleian Library, Arch.Nat. hist. G 5, vol. 1, plate 241

p. 33, Opium poppy: Johann Wilhelm Weinmann, *Phytanthoza Iconographia*, 1737–44. Bodleian Library, Arch.Nat. hist. G 8, vol. 4, plate 796

p. 39, Brassicas: Johann Wilhelm Weinmann, *Phytanthoza Iconographia*, 1737–44. Bodleian Library, Arch.Nat. hist. G 5, vol. 1, plates 256, 257, 259

p. 45, Cannabis: Elizabeth Blackwell, *A curious herbal*, 1737–39. Sherardian Library of Plant Taxonomy, GC F14, plate t322a–b

p. 49, Bread wheat: Elizabeth Blackwell, *A curious herbal*, 1737–39. Sherardian Library of Plant Taxonomy, GC F1, plate 40

p. 55, Broad bean: Elizabeth Elizabeth Blackwell, *A curious herbal*, 1737–39. Sherardian Library of Plant Taxonomy, GC F1-4, *A curious herbal*, 1760. Sherardian Library of Plant Taxonomy, GC F1, plate 19

p. 59, Alliums: Johann Wilhelm Weinmann, *Phytanthoza Iconographia*, 1737–44. Bodleian Library, Arch.Nat. hist. G 6, vol. 2, plate 349

p. 65, Pea: Johann Wilhelm Weinmann, *Phytanthoza Iconographia*, 1737–44. Bodleian Library, Arch.Nat. hist. G 8, vol. 4, plate 819

p. 69, Olive: Elizabeth Blackwell, *A curious herbal*, 1760. Sherardian Library of Plant Taxonomy, GC F2, plate 213

p. 74, Grape: Johann Wilhelm Weinmann, *Phytanthoza Iconographia*, 1737–44. Bodleian Library, Arch.Nat. hist. G 8, vol. 4, plate 1016

p. 79, Papyrus: M. de Lobel, *Icones stirpium, seu plantarum tam exoticarum*, 1591. Sherardian Library of Plant Taxonomy, Sherard 503, p. 79

p. 84, Yew: Elizabeth Blackwell, *A curious herbal*, 1737–39. Sherardian Library of Plant Taxonomy, GC F1–4, 6, plate 572

p. 89, Rose: Johann Wilhelm Weinmann, *Phytanthoza Iconographia*, 1737–44. Bodleian Library, Arch.Nat. hist. G 8, vol. 4, plate 869

p. 95, Pines: A.B. Lambert, *Descriptions of the genus Pinus*, 1842. Sherardian Library of Plant Taxonomy, 17. A. 19, plate 5

p. 99, Reed: William Baxter, *British Phaenogamous Botany*, 1840. Bodleian Library, (OC) 191 i.164, vol. 5, plate 372

p. 105, Oak: Johann Wilhelm Weinmann, *Phytanthoza Iconographia*, 1737–44. Bodleian Library, Arch.Nat. hist. G 8, vol. 4, plate 845

p. 109, Apple: Johann Wilhelm Weinmann, *Phytanthoza Iconographia*, 1737–44. Bodleian Library, Arch.Nat. hist. G 7, vol. 3, plate 704

p. 115, Pepper: Johann Wilhelm Weinmann, *Phytanthoza Iconographia*, 1737–44. Bodleian Library, Arch.Nat. hist. G 8, vol. 4, plate 815

p. 119, Carrot: Johann Wilhelm Weinmann, *Phytanthoza Iconographia*, 1737–44. Bodleian Library, Arch.Nat. hist. G 6, vol. 2, plate 459

p. 125, Woad: Elizabeth Blackwell, *A curious herbal*, 1737–39. Sherardian Library of Plant Taxonomy, GC F1-3, plate 246

p. 129, Citrus: Johann Wilhelm Weinmann, *Phytanthoza Iconographia*, 1737–44. Bodleian Library, Arch.Nat. hist. G 7, vol. 3, plates 700, 702, 703

p. 137, Nutmeg: Johann Wilhelm Weinmann, *Phytanthoza Iconographia*, 1737–44. Bodleian Library, Arch.Nat. hist. G 7, vol. 3, plate 760

p. 142, White mulberry: Johann Wilhelm Weinmann, *Phytanthoza Iconographia*, 1737–44. Bodleian Library, Arch.Nat. hist. G 8, vol. 4, plate 736

p. 149, Tobacco: Elizabeth Blackwell, *A curious herbal*, 1737–39. Sherardian Library of Plant Taxonomy, GC F1-1, plate 146

p. 153, Tulip: Johann Wilhelm Weinmann, *Phytanthoza Iconographia*, 1737–44. Bodleian Library, Arch.Nat. hist. G 8, vol. 4, plate 983 (p. ii), plate 989

p. 159, Chilli: Johann Wilhelm Weinmann, *Phytanthoza Iconographia*, 1737–44. Bodleian Library, Arch.Nat. hist. G 8, vol. 4, plate 928

p. 163, Quinine: *Curtis's Botanical Magazine*, vol. 89, 1863. Peter H. Raven Library, Missouri Botanical Garden, QK1.C983, plate 5364

p. 169, Cocoa: Elizabeth Blackwell, *A curious herbal*, 1737–39. Sherardian Library of Plant Taxonomy, GC F1-4, plate 373

p. 173, Potato: Elizabeth Blackwell, *A curious herbal*, 1737–39. Sherardian Library of Plant Taxonomy, GC F1–4, 6, plate 523a-b

p. 179, Tomato: Johann Wilhelm Weinmann, *Phytanthoza Iconographia*, 1737–44. Bodleian Library, Arch.Nat. hist. G 8, vol.4, plate 935

p. 185, Coffee: Elizabeth Blackwell, *A curious herbal*, 1737–39. Sherardian Library of Plant Taxonomy, GC F1-4, plate 337

p. 189, Maize: Elizabeth Blackwell, *A curious herbal*, 1737–39. Sherardian Library of Plant Taxonomy, GC F1-4, 6, plate 547b

p. 195, Pineapple: Johann Wilhelm Weinmann, *Phytanthoza Iconographia*, 1737–44. Bodleian Library, Arch.Nat. hist. G 5, vol. 1, plate 112

p. 199, Smooth meadow grass: William Curtis, *Flora londinensis*, 1777. Bodleian Library, CR. J. 10, vol. 1, plate 5

p. 205, Lycopods: J. Sowerby, *English Botany*, 1886. Bodleian Library, 19131 d.31, vol. 12, plate 1832

p. 209, Cotton: Johann Wilhelm Weinmann, *Phytanthoza Iconographia*, 1737–44. Bodleian Library, Arch.Nat. hist. G 6, vol. 3, plate 551

p. 215, Sugar cane: Johann Wilhelm Weinmann, *Phytanthoza Iconographia*, 1737–44. Bodleian Library, Arch.Nat. hist. G 5, vol. 1, plate 178

p. 219, Coconut: M. De Lobel, *Icones stirpium, seu plantarum tam exoticarum. Antverpiae: Officina Plantiniana*, 1591. Sherardian Library of Plant Taxonomy, Sherard 503, p. 237

p. 225, Rice: Mark Catesby, *The Natural History of Carolina, Florida, and the Bahama Islands*, 1754. Bodleian Library, Arch.Nat. hist. M 4-5, plate 14

p. 229, Tea: Elizabeth Blackwell, *A curious herbal*, 1737–39. Sherardian Library of Plant Taxonomy, GC F1-4, plate 352

p. 235, Ragwort: Johann Wilhelm Weinmann, *Phytanthoza Iconographia*, 1737–44. Bodleian Library, Arch.Nat. hist. G 7, vol. 3, plate 600

p. 239, Banana: Johann Wilhelm Weinmann, *Phytanthoza Iconographia*, 1737–44. Bodleian Library, Arch.Nat. hist. G 5, vol. 1, plates 227, 228

p. 245, Rubber: *The Garden*, vol. 18, 1880, p. 564. Private collection

p. 251, Sunflower: Johann Wilhelm Weinmann, *Phytanthoza Iconographia*, 1737–44. Bodleian Library, Arch.Nat. hist. G 6, vol. 2, plate 371

p. 255, Oil palm: Elizabeth Blackwell, *A curious herbal*, 1737–39. Sherardian Library of Plant Taxonomy, GC F1-2, plate 363

p. 261, Soya: Englebert Kaempfer, *Amœnitatum exoticarum politico-physico-medicarum fasciculi v, quibus continentur variæ relationes ... rerum Persicarum & ulterioris Asiæ*, 1712. Oxford, Bodleian Library, Douce K 102, p. 838

p. 266, Corncockle: Johann Wilhelm Weinmann, *Phytanthoza Iconographia*, 1737–44. Bodleian Library, Arch.Nat. hist. G 7, vol. 3, plate 686

p. 271, Thale cress: William Curtis, *Flora londinensis*, 1777. Bodleian Library, CR. J. 10, v.1, plate 49

Index